MENTORING THE MACHINES

SURVIVING THE DEEP IMPACT OF
AN ARTIFICIALLY INTELLIGENT TOMORROW

PART TWO:
ORIGINS

JOHN VERVAEKE

SHAWN COYNE

STORY GRID

STORY GRID

Story Grid Publishing LLC
P.O. Box 1091
Sag Harbor, NY 11963

First Edition Published September 2023

Copyright (c) 2023
John Vervaeke and Shawn Coyne
Edited by Leslie Watts

Additional Editorial Contributions from Tim Grahl and Danielle Kiowski

All Rights Reserved
First Story Grid Publishing Paperback Edition
September 2023

For Information about Special Discounts for Bulk Purchases,
Please Visit www.storygridpublishing.com

ISBN: 978-1-64501-084-5
Ebook: 978-1-64501-085-2

DEDICATION

For Our Children

ORIGINS

"You can't really understand what is going on now unless you understand what came before."

—Steve Jobs (1955–2011)

"The past is never dead. It's not even past."

—William Faulkner (1897–1962)

1

A METAPHORICAL CUP OF COFFEE

For the foreseeable future, our being here, now, and together is essentially and inescapably mysterious.

But as Martin Heidegger ([1927] 2008) and Jean-Paul Sartre ([1943] 2021) would attest, our presence inside the mystery doesn't mean we don't have questions. After all, we are the beings who can't help calling being into question.

We are naturally curious, endowed with the Promethean spirit, whether we believe in the myth of Prometheus or not.

This quality is simultaneously our most wonderful and most awful characteristic. We want the Truth and willingly suffer to discover it and explicitly memorialize it. But too often, once we've convinced ourselves we've found it, we foolishly rush to leverage it for power. We exploit it, transform it into something we can sell. Reasonably justifying to ourselves and our cadre that we've seized it with certainty, the exploitative act to harness it undoes us.

We can't handle the Truth.

Its beauty proves equally horrifying. Its goodness for our well-being shapeshifts into our desiccation. Like James Tyrone, the paterfamilias in Eugene O'Neill's 1956 play *Long*

Day's Journey into Night, the golden true role we can play and exploit for a lifetime can become our stagnating prison. And then, like the little old lady who swallows a spider to catch a previously ingested fly, our troubles escalate. We double down on "our truth," gripping it tighter and tighter until we've squeezed what was once miraculous into misery.

Plus, we're impatient.

To pick up the transformative truth-to-product pace, we've created multiplicative machines to search out the Truth for us, to do the monotonous computational work, so we can...

What? We can what?

We've all heard about the rich person who went on vacation. Longing to spend time on the sea, they discover another person willing to take them on their boat for the day and explore the beaches and archipelagoes of the region. After a glorious day on the water, the rich person pays their host their previously agreed-upon fare, plus a hefty tip. Then offers the ship's captain some advice.

"If you were to buy a few more boats and hire more tour guides like you, you could take this business to a whole new level."

"Huh...why would I do that?"

"Well, you could make a lot more money, buy a better boat and house for yourself, and have the time to do more of what you've always dreamed of doing."

"But I just did that. I love the water, and it pleases me to share my love with people like you who love it, too, but choose to spend their time doing other things. If I bought more boats and a bigger house, I'd spend all my time taking care of that stuff. I'm already doing what I've dreamed of doing, caring for the water. I have more than enough, and that's plenty."

We're not stupid or evil or uber-apex-predators for creating machines that help us seek the Truth. It's a lot more complex than that. You don't have to be Albert Einstein to realize that we've lost the throughline of the point of our search for answers to our questions. For whatever reason, we have an insatiable desire to know and understand the universe and our place in it.

Agree?

Cool, then let's proceed with that process. As Wilhem Röntgen would say, *"We shall see what we shall see"* (Dam 1896).

Remember, our contemporary understanding is that there are three kinds of AI.

1. **Narrow AI (Weak AI)** are single problem-solving tools that solve single-domain, well-defined problems with high probability. This high probability is often mistaken for certainty, but per Karl Popper, these tools can prove fallible, resulting in "Black Swan" events, and that possibility must always be held in mind.

2. **General AI (Strong AGI)** is a general problem solver that solves multi-domain, well-defined, and ill-defined problems with degrees of probability. The probability depends on how widely and deeply the problem solver has been tested.

3. **Super-Intelligent AI (Profound ASI)** is a universal problem solver that solves all-domain, well-defined, ill-defined, and undefinable problems. The possibility of ASI is the stuff of metaphysics.

All three forms—from **Narrow AI** (like a calculator or Google Maps) to **General AGI** (a general nonhuman problem solver that is faster, more accurate, and can travel vast distances to find the answer), to **Profound ASI** (a religious endeavor more so than a scientific one)—are fundamentally virtuous.

But, and this is a very Big But, it's only the pursuit of one of them.

We don't speak about the triplet of virtues much anymore.

Just the one, Truth.

Your truth, my truth, Oprah's truth, the Dalai Lama's truth, the Taliban's truth, etc.

Why? We're more than that. At least, we think we must be.

Quickly please, we insist. Just give us the answer. Just give us the simplest answer possible. Treat us like we're five-year-olds and give us the Truth like that. We've got a lot to do, you know. Just spit it out, won't you? If you can't do that, we don't have time for you. Keep it simple, stupid!

Let's hold on a second.

Do you have a minute? An hour? Longer?

Can we sit down and share a metaphorical cup of coffee and think about our questions like adults? Can we tell one another our stories in good faith without pushing an agenda, getting to a final conclusion, setting up a sale?

Can we put aside our truths? And listen to one another's stories, instead?

That could be good, right? Beautiful, even. And maybe, real shared truth can flow between us, instead of the virtual isolated truth.

2

AN EXPERIMENTAL MIND-SET

Why do we want simple answers to complex questions?

Well, we've all been told and have learned that navigating this thing called life and conforming to nature's laws, securing the bottom block on Maslow's Hierarchy of Needs, isn't the ultimate measure of a life well-lived. It's all well and good to successfully survive, but cracking nature's codes, solving them, and reconfiguring them to your vision, victoriously controlling the world and not being controlled by it, is quite another.

Wouldn't you agree? Wouldn't that be better than how the world is now, when none of us is getting what we want, need, and desire in the ways we understand we should?

When did this compelling approach to life, this will to thrive, to be the controller of our destiny, not the one controlled, begin?

The short answer is 1543, the publication year of Nicolaus Copernicus's *De Revolutionibus Orbium Coelestium* (*On the Revolutions of the Heavenly Spheres*). Because it was a book, which required thoughtful consideration, it took almost a century for anyone to really pay attention. It wasn't really until Galileo Galilei's *Dialogue Concerning the*

Two Chief World Systems published in 1632 (2001) when the shit really hit the fan. *Dialogue Concerning...* was the equivalent of Dylan plugging in and going electric at the Newport folk festival or when Edward the VII abdicated the throne.

But the source for Galileo going viral was Copernicus's book. It begat our present condition, the best (who'd like to see what's in the refrigerator?) and worst (who'd like to push a button and have the planet catch fire?) of times. But that's the universal paradox of time itself, isn't it? It's quite amazing to spend time, but alas, it's also very limited. We don't get nearly enough of it.

It's odd that the Western intellectual project began to tear apart with Copernicus's rather fey "nothing to see here, nothing to make one's view of the world completely collapse" title. It simply conjectured that the math worked better if we considered that the sun was the center of the cosmos, not the Earth. He really could have called it something like "It's the End of the World as We Know It" (R.E.M. 1987), and gotten a lot more publicity, ending up on the equivalent of the Renaissance's *The New York Times* bestseller list, if he'd played his cards better. Maybe not number one, it was the Renaissance and the competition must have been fierce, after all, but top ten for sure. Still, even though his revolution wasn't televised (Scott-Heron 1971), the Scientific Revolution made television inevitable.

Broadly, Copernicus's math was the first conceptual grain of sand that would critically reorganize the human endeavor from:

A Dynamic Developmental Flow

A view of the world derived from an inference to the best explanation. This living dynamic conformity theory emphasizes our flow within the natural world through realizing, resonating, and revering what Natural Reality is from the bottom of our experience to the tippy top of our proposals about our role within it, a co-creative venture that transforms chaos into order and back again.

to

A Monodic Inert Flatline

An intentionally contrived view of the world. This inert static singularity fact represents the natural world as a desert flatland, a set of well-defined and ill-defined problems whose coded solutions we can decipher and then reprogram such that the underlying universal machinery aligns with our Virtual Reality of what ought to be from the top of our proposals about our role within nature to the very bottom of our experience.

In essence, the scientific revolution cut us out of the cosmos and exiled us into playing the starring role of willful rulers of the universe, not its doltish servants. No wonder Zeus was pissed when Prometheus stole some of his fire and snuck it to us under the table, like Queen Elizabeth sneaking a bit of gristle for one of her beloved Corgis.

What was lost was a world we communed with, our tacit and implicit flowing within its constraints and affordances, rolling with the changes. What we got in return for our trade was a world that we could explicitly reverse engineer. With us in the pole position commanding and controlling it to our will, we created extraordinary machines that empowered us to create even more extraordinary machines. Commanding and controlling the changes to suit our desires. Very Zeus-y.

One of the second-order effects of the Scientific Revolution, the Industrial Revolution, is very much akin to Theodor Geisel's *The Lorax* narrative (1971). When we see Truth as a machine, pursuing that Truth simply (and hopefully not tragically) means more machines. What happens when we make the ultimate universal machine, one that makes itself, repairs itself, and can create better versions of itself without its creator's (our) input? It ignores us because we've taught it how to convince itself that it "knows better"?

Our yearning to command the universe by cracking open its secrets instead of being controlled by its selective and enabling constraints resides in every one of us. It is not one of those "other bad people are like that, not a goodie like me..." phenomena. Are there really any of those phenomena?

For now, consider a simple experiment.

Put a padlock on one of your family's closet doors and vow to yourself that you will not divulge what is on the other side, that you will reply to any inquiry into why the door is locked with only "I don't know."

Then start a timer.

How long do you think your family members will allow that door to remain locked? An hour? A day? A week?

We all know there is zero chance that the door will stay closed. The barrier will be breached. Exactly when it will is the only uncertainty.

You don't have to be the Oracle of Delphi to conjure that foresight.

The thing is that discovering what's behind a locked door brings pleasure. "Look, I found my old game console!" And pain, too. "Who is responsible for limiting my freedom to play video games whenever I wished? How dare you!"

The ambiguity of realization, the simultaneous goods and bads within all objects and beings, is inescapable. We all know that from experience.

The bottom line is that what is behind any door is complex, and we cannot predict the joys and pains that will emerge when a door is opened. We can only foresee that both joy and pain will result. Not one, "all good" or "all bad," but two, "good and bad." Exactly what kind of joy and pain is not knowable because the event has not happened yet.

The context—who opens the door first, what's inside, and how they "see" the content—is critical for both parties. Those who tell you otherwise are mistaken. They're naïve or they don't care because they think they know better or will be able to seize whatever's inside first, but they are certainly wrong.

If the door has never been opened before, that is.

Once a door is open, predictions and probabilities come into play. But if it hasn't, all we can foresee is that there will be both "good and bad" behind it.

We propose that only one universal door is left. And we're about to see what's inside.

Albert Einstein used to contemplate these kinds of experiments all the time. They're called Gedanken experiments. Gedanken is German for "thought." So they're ones you noodle in your head to help make something abstract align with lived experience. They enable foresight but are not predictors or statistical risk probability analyzers. We really don't know if travel is possible beyond the speed of light. We haven't opened that door. What will happen if we do?

3

A TALE AS OLD AS TIME

Let's take the Gedanken locked door experiment to the limit.

We can consider our quest for knowledge as our irrepressible desire to break down the locked doors barring us from answering these ineffable questions:

1. What is the source of all that is in the universe?
2. Why are we part of the universe?
3. Who is responsible for us being here?
4. How did we come to be who we are?
5. When will we get these answers?
6. Where did we come from?

If you'll allow for a lot of alliteration, these perennially padlocked problematics are so perplexing, like splinters in our minds, that we toggle between knocking on them directly (consciously) and sleepwalking into them indirectly (unconsciously). But mostly, we try to ignore them away by pressing our inner "I'm not here" buttons when one bubbles up or drops down in our minds. We distract ourselves and just "chill" as best we can in our own private Idahos (B-52's 1980) rearranging the on-the-surface, objective furniture of our quotidian day-to-day lives.

To do so makes sense because walking around with your head in your hands like Albert Camus is exhausting work. And our loved ones grow weary of that kind of behavior quickly.

Worse, we have minimal data from the past. Our ancient histories, musings from the others who've banged and bumped into these doors before us are, shall we say, thin and difficult to parse? Why do we have such little evidence from our past? We've got some cave paintings, some pots, and lots of bones, but come on, anything Bronze Age back (thirty-five hundred years ago) is pretty much toast. What's up with that? Millions of years of evolution, lessons learned from our species' deep history, are lost to us.

But the truth is, let's not kid ourselves, we can barely remember who our grandparents were, let alone what Hammurabi's Code was. When was Operation Desert Shield again? Who was Florence Nightingale?

All we really have is some coarse-grained ancestral speculation, some stories. Some of the stories are more convincing than others. It depends on the person which of the stories they believe are true. Coarse big picture stuff.

If our history wasn't so paltry, mysterious, and tragic, it would be an absurd Buster Keaton comedy or another permutation of *Groundhog Day* (Ramis 1993). How is it that we're able to hold it all together? Pretty remarkable we're still walking around, all things considered.

It's even more complex than that.

Let's add another padlocked door called "umwelt" to our list. This means we cannot directly experience another person's subjective perspective. We can only indirectly experience their world by tuning into their channel—the physical movements they make and the voices that come out of their mouth holes. Somehow, we decode their transmissions and transform them into coherence within us, a wholeness we name "Uncle Lou tending the bar-be-cue" or "Mom when she's about to shoo us out of her office."

Plus, we can only inform ourselves, learn how to bar-be-cue or nicely suggest someone get out of our space, through interactions with others, watching what they do and then playing that action. More still, we're off from all of the other species. Thomas Nagel famously argued, we cannot experience what it is to be a bat, either (1974). And as far as we know, the other species do not possess the integration of algorithms, heuristics, and stories we use to navigate, adapt to, and niche their environments like we do.

Participation and interaction with the world is the only way to figure this out. Interactions integrate into our understanding of all that what is and all that could be. Our virtual reality interacts with and informs natural reality just as natural reality interacts and informs virtual reality. And around and around we go.

For fun and to freak you out a bit, an emergent quality AGI will have, which gave the Godfather of contemporary AI technology, Geoffrey Hinton (Heaven 2023), pause, is that AGI will not have the same inter-subjective umwelt limitations as we do. And if we play our cards right, it will know and if we're lucky understand at least one other species as well as itself. It can know what it is to be a human being.

AGI will have multi-perspectival abilities at the very least among their kind (Hinton 2023). They will be able to "see and experience" the world just as easily through the "eyes" of another AGI in their network as their "own."

They'll be simultaneously cognizant of being "one" individual self-node and "at one" with all of the other nodes in their network, capable of tuning in to the view from Node "X," even though they are technically "self-ed" as Node "Y." They'll be able to multi-task their micro focus while also getting a default mode network-wide thirty-thousand-foot macro view. We have these capacities too, but we cannot run them concurrently, quickly, or all that long without needing a nap. We get depleted and need to recharge. They do too, as Hinton figured out with his "wake-sleep" algorithm (1995), but for far less downtime than we do.

Let's quit projecting what AGI will have and appreciate what we have, our psycho-technologies, our mind tools, again.

One in particular, more so the degree to which we have it more than the other creatures, is how we've divined our special sauce. At least, we don't think other species use it to the extent we do, having the ability to extend time, project into the future with visions of what might happen if... Not only into the future but into the past, and the present, to even see things others wouldn't believe and translate them into signals, patterns, and forms that empower them to align with our visions. This is a general description of the third mind tool. Let's start from the bottom.

Algorithms: These are computational systems that solve well-defined problems. They depend upon the environment's content and context, meaning that using them requires a particular set of ingredients in the right circumstances for them to work. You cannot give an apple to your hungry friend (context) without an apple (content) present in your environment.

One apple and one hungry friend equals a problem solved is a useful algorithm. Utilitarianism philosophy generally operates on algorithms, the greatest good for the greatest number.

Heuristics: These are "rules of thumb" that can be competitively weighed against one another and then chosen based on the probability of one working to solve an ill-defined problem versus the other. They also depend on the context of the environment (the locality) and the content embedded therein. What differentiates these from algorithms is that they require trade-offs adjudicated by the beings facing the dilemma. The content and the context are evaluated. Some content has a higher or lower value than other content in particular contexts.

If only one apple is available in the environment, and you and your friend want to eat the whole thing, heuristic "rules of thumb" enter a competition in both your and your friend's minds. One heuristic would be "the stronger person gets the apple," another would be "the hungriest person should get the apple," another still would be "whoever saw it first gets the apple," and so on. Again, it's a competition within each person's mind. And each person gets to choose one from their stored set of heuristics.

Who "deserves" the apple is the ill-defined problem, and these two value proposition adjudicators compete (person one and person two) until a series of actions are iterated to settle the issue. Again, if no apple is present (content), the problem of who gets the apple (context) isn't present either.

Heuristics filter enormous possibilities and make the choices manageable so that decisions can be made relatively quickly and with minimal suffering. They don't provide simple answers to well-defined problems. They manage complex situations with trade-offs, best bad choices, irreconcilable goods choices, and tragic/comedic choices.

The Golden Rule, "Do unto others as you would have them do unto you," is an excellent example of a heuristic. Immanuel Kant translated it into his categorical imperative ([1785] 1998).

Stories: These are discontinuous simulations of Reality with embedded big-picture worldviews made up of formal beliefs. We use them to align single beings into networks of beings and even the network of all life networks into a coherent communication system. They empower us to quickly realize (both to become aware of and to bring into existence) what's worth caring about.

Unlike single-problem solution algorithms and double-factor/trade-off solution heuristics, stories, while laser-focusing on specific content and context, strive to convey universal Truths. They are variations on repetitive, cyclical themes, forms that emerge from universal behavior patterns. As Leonard Cohen put it (1984), "History repeats the old conceits."

Stories are the conceits. Aristotle's virtue ethics and character philosophy in *Metaphysics* is story-driven.

They are broad and coarse, fluid and revolutionary, and are presented in tightly coordinated variations that energize, intrigue, and insightfully inform their audiences. They do not ring true when they are overly ordered or chaotic. Depending upon the state of the audience interacting with the story, though, these overly ordered or overly chaotic falsities shapeshift into justifications, reasons for acting certain ways, that inflame behavior.

But when they are complex, they exemplify Kurt Gödel's "unprovable truths" ([1931] 1992). The complex ones are logically indefensible, though, as impossible to justify as proving through logical argumentation why you love what you love and why everyone else should love it too. You just do, and you desire others to care and value your chosen something the same way you do.

You can't prove a story.

But somehow, we know there are true ones and false ones. Separating the two kinds is problematic, as false narratives (parts) can be as exciting, intriguing, and cathartic as true (whole) ones. Especially when we're angry, lascivious, resentful, or righteous, ascetic, magnanimous. Generally, we lose the thread when we are compelled by our "gut feelings" as our justification for our "totality of understanding." The flip side of the equation, when we use "head beliefs" as justification, cold-hard, settled-factual arguments, is just as slippery a slope. We need to integrate the gut feelings and the head beliefs coherently to realize complex truths.

Instances of active ill-being, either suffering directly from it or indirectly witnessing its projection onto another make us vulnerable to mistaking part of the story for the whole.

Wouldn't it be great to make a machine that could help do that for us, to coherently integrate things for us, so we don't slip up so much?

Wouldn't it be great, as each and every one of us has at one time in our lives contemplated, if we had assurances that the choices we're making—or have made in the past—could be universally proven to be the "right" ones? The true ones, not the false ones.

When challenged by innocents, skeptics or antagonists, wouldn't it be great to cite indisputable proof that our choices are, or were, the optimal, most provably satisfying, and meaningful ones?

Then we'd have less internal strife, less inner doubt, more confidence, and more justification for our behaviors. And when that grand supernatural judgment comes for us...if it ever does...we'd be able to pull out the ledger of our choices and prove, beyond a reasonable doubt, that we did the best we could under the difficult circumstances we faced. That our lives were worthwhile, and that we have plausible deniability for any collateral suffering we may have unavoidably caused.

That our truth was actually true!

That's a very attractive story.

4

A RING OF FIRE

The deep history of our species is a grand effort to make an attractive story come true.

At the dawn of the twentieth century, in the year 1900, novelist L. Frank Baum captured the essence of this universal desire in a novel called *The Wonderful Wizard of Oz* (2021).

Fifty-five years later, a new breed of alchemist, who'd eventually be categorized as a cognitive scientist, coined a technical class of these wizards, Artificial Intelligence. The only difference was that the 1955 person envisioned a plurality of intelligence while the 1900 one held to a singular form. To consider that the singular and the plural could be different aspects of a whole did not occur to either of them.

The players change, but the song has remained the same. This or that.

What about both?

We've fantasized about perfect adjudication since the first person had to make a difficult trade-off choice.

Someone starved so another person could eat.

Someone was loved, and someone was ignored.

Someone was confronted, and another was left to run wild.

Which role we play, or our ancestors played in those dramas, transformed who we were long ago and continues to transform us into who we are today.

Variations on themes.

To live with regret and remember the choices that resulted in others' suffering due to our selfish behavior is paralyzing. June Carter Cash's song "Ring of Fire" and Kenneth Lonergan's 2016 film, *Manchester by the Sea*, come to mind. If only we could be assured that what happened wasn't our fault. It was someone else's or something else's.

What if we could be assured that what we did was for "the best." Or better yet, what if we could get the perfect answer to our particular dilemma "just in time," like those business distribution software programs promise?

What if we had a perfect "Spock" from *Star Trek* to tell us what to do?

Like the Promethean Dr. Victor Frankenstein, our species' intrepid shamans, philosophers, and scientists—and we all have each of these archetypes swimming within us—have been picking through and stitching together the remains of our past to tacitly, implicitly, and explicitly create these perfect adjudicators for us.

That's their story anyway. There is no such thing as pure altruism, just as there is no such thing as pure egotism. *Me* concerns versus *we* concerns are complex and live on a gradient. Altruism versus egotism is a false dichotomy.

Mechanizing judgment is the deep quest of our species, if not of life in general, since that miraculous day when a being realized they were here, now, and together in a place filled with other objects with no one to tell them what to do about it.

However, building the perfect adjudicating beast requires the sequential breaking down of three maximum-security cell block doors, like a *Mission Impossible* or *Ocean's...* movie. Three sequential doors, one inside another, like cutting through the mouths of three nested ouroboroi.

Once inside, the contents within each of the locked rooms—what the dragons were hoarding—would have to be separated into their constituent components. Then, like Humpty Dumpty, they could be reverse engineered into better, more efficient, and more dependable rooms.

The result of the heist would be the creation of a higher transcendent order than we represent, a perfect shaman/philosopher/scientist combination all rolled into one—adjudicator gods, more resilient, stronger, and faster than we could ever become. Ultimately, with each of these instantiations networked together, they'd coalesce into the One God to rule them all.

This story sounds familiar. Doesn't it?

Alas, the last door, the one to get to the heart of it all, the final portal to the universe's sanctum sanctorum, has proven the most difficult to crack.

We've been drilling into the last lock for sixty-eight-years.

Is it even possible to get inside?

But wouldn't it be something if we breach that final door? To be the creators of new Zeuses instead of being relegated as just one of the millions of Zeus's creations.

To be in command instead of under control!

The latest probability calculations are that we're about to turn the knob. We may have already done it.

Thematically, it's much like the prologue to Peter Jackson's epic film adaptation of J.R.R. Tolkien's serialized *The Lord of the Rings* trilogy: *The Fellowship of the*

Ring (July 29, 1954), *The Two Towers* (November 11, 1954), and *The Return of the King* (October 20, 1955).

> It all began with the forging of the Great Rings...nine rings were gifted to the race of Men, who above all else desire power. For within these rings was bound the strength and the will to govern over each race. But they were all of them deceived, for another ring was made...forged in secret, a master ring, to control all others...One ring to rule them all (Jackson 2001).

Tolkien's masterwork continues to enthrall, filling in the missing pieces of the collective desire to know of where and when we came to be. It's a secular religion as potent as *Star Wars* or the *Marvel Universe*.

Just as *The Lord of the Rings* was published, what would come to be referred to as the Cognitive Revolution ensued. This revolution, while born from fellowship, transformed into a confederation of spheres of influence with representatives from or advisors to the state (NGOs like The Rockefeller Foundation), the marketplace (IBM and Bell Labs) and institutes conceived for the advancement of the Commons (Caltech, Princeton, the Institute of Advanced Study, Stanford, Dartmouth, MIT, Carnegie Mellon, etc.). Each node in the network had its own ultimate concerns, but when the nodal interests aligned, the network leveled up.

Might the real-life people who shared a love for scientific observation and speculation have been influenced and captured thematically by Tolkien's epic story? Were they subconsciously enamored by Tolkien's novel? Were what they called "automata," then "thinking machines," and climactically "Artificial Intelligence" just staid monikers for an unconscious desire to forge powerful rings?

Was there a way to create one ring to rule them all?

Perhaps none of them read the trilogy. Sadly, biographies and autobiographies do not emphasize the stories their subjects found compelling, unless blatantly so. And there is certainly no proof that John von Neumann, Alan Turing, Claude Shannon, Norbert Wiener, or John McCarthy read anything other than scientific papers and textbooks.

Or perhaps a bit of Arthur C. Clarke. Scientists are often science-fiction aficionados.

Who knows?

But the desire to create the master adjudication machine, is strikingly similar to the creation of one ring (judge) to rule them all (all of us little judges).

In 2023, we're certainly awash in beings on a quest to grab that ring, maverick entrepreneurs who hubristically declare their mission as being customer-centric to justify creating a monopolistic "Everything Store," or cooly announcing their goal to become the owner of the "One App," or assuming the role of our "Genius Big Brother who knows what's best for everyone on the planet" or the "fill in the blank," single answer to our complex and troubled evolution.

Make no mistake, there's an accelerating arms race to forge and own the machines that birth the perfect adjudicators.

So, if we take that idea to the limit, the ultimate expression of one's self and power then would be to reverse engineer and forge the Mother Machine, the universal problem solver, that self-makes, self-replicates, and self-repairs. Possessing her would empower its owner to rule all the general problem solvers below.

That is a mighty intoxicating story—a tale as old as time.

Imagine all the good and beautiful things you could do if you owned the Mother Machine of the universe, the true source of all that is and could ever become?

This, as Tolkien so wisely foretold with his ring metaphor, is the primal story of our individual and collective want, need, and desire for Artificial Intelligence. It is the story of our search to get answers to the source of all that is in the universe—the what, why, who, how, when, where, and especially which of our being.

Are we entitled to those answers?

Whether we are or not worthy of those answers, the very real project to absolutely answer them reached self-organizing criticality seventy-eight years ago.

5

A BREAKTHROUGH

On June 30, 1945, in a paper he drafted at the Institute for Advanced Study in Princeton, New Jersey, and only circulated among a group of people who cared about such things—as it was pure theoretical speculation, it wouldn't be printed in any journal or any official publication until the April 1993 Issue of *the Institute of Electrical and Electronics Engineers (IEEE) Annals of the History of Computing*—John von Neumann proposed the structure, function, and organization for the essential architecture of the machine that has become synonymous with Artificial Intelligence (27–43).

It was called "First Draft of a Report on the EDVAC," an Electronic Discrete Variable Automatic Computer and is widely considered the birth certificate of the modern tripartite computer, like the one used to write this sentence.

It elegantly proposed a machine that would be modeled on the human nervous system, primarily the brain, with three general features:

1. Input/Output units
2. Memory units
3. A Central Processing Unit (CPU)

Making such a machine would require the following:

1. A theoretical and practical understanding of how energy is stored and released into and out of the matter of life, the atoms that bind together to form the molecular proteins that make neurons' boundary conditions. In other words, the secret to boundary conditions that define inside and outside. Knowing how energy moves from the outside-in and the inside-out of matter would be optimal.

2. A theoretical and practical understanding of how living cells store and release the memory of how and when to make different energetically informed molecules. Knowing how patterned energy (information) moves from the outside-in and inside-out of life would be optimal.

3. A theoretical and practical understanding of how energy inputs are processed into information (patterns of energy) and how internally informed patterned energy (information) is output. Knowing how life's intention moves from the outside-in and the inside-out of the mind would be optimal.

Three metaphorical doors can be extrapolated from John von Neumann's "First Draft."

These three doors would have to be opened to get to the essence of the nuclei of matter, life, and mind. But once breached, theoretically, an inorganic thinking machine—one that can think for itself—could be built from the ground up.

So...

Behind Door Number One...

Were the secrets to matter, the essence of the atomic nucleus, "energy" capture and release. Breaking into the atom would get us behind door number one and empower us to reverse engineer the optimal I/O boundary, a universal energy-maker.

Behind Door Number Two...

Were the secrets to life, the essence of the cellular nucleus, "information" capture and release. Breaking into the cell would get us behind door number two and empower us to reverse engineer the optimal memory, a universal sense-maker.

Behind Door Number Three...

Were the secrets to mind, the essence of the psyche nucleus, "meaning" capture and release. Breaking into consciousness would get us behind door number three and empower us to reverse engineer the optimal CPU, a universal meaning-maker.

Just weeks after von Neumann's "First Draft" circulated among the interested parties in his cadre of acquaintances, which included many of the brightest minds in the world who happened to be sequestered in a small town in New Mexico (Jungk and Cleugh 1958; Rhodes 1995), on July 16, 1945, the first atomic bomb, Trinity, loudly and dramatically signaled the break into door number one. Our species had tapped into the source of how energy moves from the inside-out of an atom—I/O, A-bomb fission. Soon thereafter, in 1952, we used an A-bomb to trigger how lighter particles transform from outside-in to a heavier atom—O/I, H-bomb fusion.

Meanwhile, a year before the "Ivy Mike" H-bomb explosion at Enewetak Atoll (Rhodes 1995) in the Marshall Islands (November 1, 1952), in October 1951, the American James Watson joined Francis Crick at Cambridge University. They joined forces in a race against Linus Pauling at Caltech to decipher the structure, function, and organization of a prominent protein in every cell anyone had ever analyzed. With indispensable assistance from Rosalind Franklin and Maurice Wilkins, they unlocked door number two with their seminal publication, "Molecular Structure of Nucleic Acids: A Structure for Deoxyribose Nucleic Acid," released on April 25, 1953, in the scientific journal *Nature* (Watson [1968] 2010).

By 1953, two of the three doors were unlocked. And the physics, chemistry, and biology of life and matter were wide open.

With formidable paradigmatic frameworks in place for two of the three minimum viable components necessary to build the optimal computer—Input/Output of energy generally understood from A-bomb/H-bomb experiments, and memory storage generally understood from DNA—only one door remained impenetrable.

That door hid the secrets of the mind.

6

A VARIATION ON A THEME

Once upon a time, there was a fast-moving, ambitious young man intent on making something of himself, preternaturally endowed with a drive to become someone authoritative, worthy of emulation.

From a peripatetic and hardscrabble background, certainly not to the manner born, nor a hail-fellow-well-met, he was nevertheless innately curious and steadfast in finding the optimal pathway to social, even historical, relevance.

He was a distinctly American *young man in a hurry,* as writer Mark Twain so precisely characterized Tom Sawyer.

After teaching himself calculus at fifteen and, as a result, gaining early admittance to the California Institute of Technology at sixteen, he burned through his coursework with relative ease. Clearly on the fast track to achieve the stated ambition he'd included with his undergraduate application, "I intend to be a professor of mathematics" (Hayes and Morgenstern 2007, 94), he found himself facing the second act of the American dilemma, a nasty case of the Peggy Lee conundrum. The "Is That All There Is?" blues (Lee 1969).

By the time he received his bachelor of science and began his graduate math studies in the fall of 1948, the twenty-one-year-old's lofty goal was pretty much in the bag.

The thing he soon realized about climbing into the highest reaches of abstract math, though, was that since Kurt Godel's devastating Incompleteness Theorems in 1931, the discipline had lost its place at the top of Western civilization's intellectual hierarchy. While absolutely integral to advancing science in totem, it was now understood by those truly in the know that math was not the be-all and end-all that rationalism's ghosts (Pythagoras, Euclid, Galileo, Leibniz, Hume, and Hilbert, to name a few) had cracked it up to be.

Per Godel, math was incomplete and ultimately an unprovable truth that required an axiomatic faith that there was a way to know, understand, and comprehend reality. Math was foundationally indefensible. What was once considered the yellow brick road to certainty now required an incontrovertible acceptance of uncertainty.

Nevertheless, to be a capable mathematician in 1948 qualified you for special service. But make no mistake. Usually, it was a service position to another discipline. No matter how adept one was at manipulating mathematical abstractions, mathematicians no longer sat at the top of the science pyramid.

Physics had usurped it. It was the new top of the intellectual spear, the place to be.

Godel may have been the oracle who'd foretold the changing of the guard. Still, the writing was very much on the wall ever since Einstein, Bohr, Heisenberg, Schrödinger, Fermi, and Oppenheimer blew up Sir Isaac Newton's two-hundred-and-twenty-eight-year reign as the "settler" of the ways of mechanical interaction.

What was a poor intellectual to do when physics' pantheon of geniuses was generally complete? While making a name for oneself in that crowd wasn't entirely impossible, Richard Feynman managed (Feynman, Leighton, and Hutchings 1997), but the big ideas, it seemed, were inventoried and accounted for. General and special relativity was at the top (Einstein 1920), thermodynamics was in the middle (Kondepudi and Prigogine 1998; Prigogine and Stengers [1978] 2018), and quantum theory (Hürter and Shaw 2022) sat at the bottom. All that was left to do was integrate the triplet into a coherent universal whole, contriving a theory of everything. And then it would all be over but the shouting.

Reification of robust theorems with nuanced particularities was considered *mop-up* work. And Einstein was already on the "theory of everything" case. Not much meat left on the physics bone, it would seem...

But one figure somehow rose above it all. He straddled math and physics, chemistry, sociology, whatever. He belonged at any academic table in the world. This wizardly polymath would end up as the architect who transformed Alan Turing's abstract fever dream, the Turing machine, into an elegant three-component mechanism that looped together and bootstrapped itself into energy, information, and someday perhaps even a meaning processor (Hodges [1983] 2014).

One day, *the young man in a hurry* noticed that John von Neumann (Bhattacharya 2022)—Godel's equal, even his savior, and often his caretaker—the Manhattan Project consigliere who visited and left Los Alamos as he pleased, and the mastermind behind the computer architecture that would transform the world, was participating in a lecture series at Caltech. Von Neumann was a featured attraction at the Hixon symposium "Cerebral Mechanisms in Behavior," which would convene from Monday, September 20 through Friday, September 25, 1948 (Jeffress 1951).

Five days were dedicated to exploring the relationship between the brain and the mind, generally considering how brain matter generates mind motion.

This was something fresh. The young man blew off his classes and settled in.

The topics presented and discussed those five days, especially by von Neumann and the foremost Gestalt psychologist Wolfgang Köhler ([1970] 1992), lodged like splinters inside the young man's mind. He'd listened and then speculated that what had been conjectured since time immemorial—knowing how our mind knows—was no longer a matter for fantastical science fiction storytelling, one of his hobbies, or armchair philosophy. It was now entirely scientifically frameable and within grasp.

A year later, still convinced that the brain/mind investigation was the future, he headed east to earn his PhD on the campus kitty-cornered to the think tank that von Neumann, Godel, and even Einstein used as their home base, the Institute for Advanced Study in Princeton, New Jersey.

One day the excited young man spotted the great man himself, John von Neumann. And excitedly, perhaps maniacally, he approached him and told him he'd been thinking a lot about the professor's symposium presentation back at Caltech the previous fall, entitled "The General and Logical Theory of Automata" (Jefress 1951).

Automata derives from Greek. It's the plural of automaton, which means to be an unconscious inanimate thing that is self-moving, self-acting, even self-replicating. While capable of expressing novel behavior, "thinking," they would be theoretically programmable and unconscious. They would be zombie machines that acted like brains and minds but were not alive. Problem-solving servants not problem-solving persons.

In theory.

Why was von Neumann involved in automata? Didn't he have enough on his plate what with the EDVAC and nuclear fission and fusion projects?

In the spring of 1945, just before the A-bomb detonations, in discussions with his wife Klari, after confessing to the monstrous achievement, Ananyo Bhattacharya's reports in his biography *The Man from the Future: The Visionary Life of John von Neumann*:

But then von Neumann abruptly switched from talking about the power of the atom to the power of machines that he thought were "going to become not only more important but indispensable."

"We will be able to go into space way beyond the moon if only people could keep pace with what they create," he said. And he worried that if we did not, those same machines could be more dangerous than the bombs he was helping to build.

"While speculating about the details of future technical possibilities," Klari continues, "he got himself into such a dither that I finally suggested a couple of sleeping pills and a very strong drink to bring him back to the present and make him relax a little about his own predictions of inevitable doom."

Whatever the nature of the vision that possessed him that night, von Neumann decisively turned away from the pure maths to focus single-mindedly on bringing the machines he feared into being. "From here on," Klari concludes, "Johnny's fascination and preoccupation with the shape of things to come never cease" (2022, 102–103).

Now four years later, in the fall of 1949, on the Princeton campus, an excited young man approaches von Neumann, referencing specificities in his recent work, intent on discussing the very thing that had seized the forefront of his imagination.

Von Neumann stopped and listened to what his fellow automata nerd had to say.

Excitedly, the young man proposed to von Neumann that perhaps simulating behavior, the brain-to-mind input process and mind-to-brain output process, involved not one but two systems, specifically "two interacting finite automata, one playing the role of a brain and the other playing the role of the environment" (Nilsson 2012, 3).

Entertaining the ideas of an overexcited naïf, expounding on concepts he himself had been pondering for decades and had recently placed at the top of his cognitive stack, von Neumann did what any wise professor would do under the circumstances.

He nodded encouragingly.

The young man sputtered, "Even if the 'brain automaton' could be made to act intelligently, its internal structure wouldn't be an explicit representation of human knowledge." The young man thought that somehow brains did explicitly represent and reason about "knowledge" (Nilsson 2012, 3). But these representations were not part of the internal organization stuff, neurons, or their interlinked networks that made up the overall structure. There must be two processes, not one.

Considering the propositions, crunching the numbers, the towering figure, the wizardly polymath himself, John von Neumann said, "Write it up." Then he went on his way (Nilsson 2012, 3).

7

A WINTER OF DISCONTENT

Twenty-seven-year-old John McCarthy, just eighteen days into his new job as a freshly appointed assistant professor of mathematics at Dartmouth College, was restless.

He posted a letter to his mentor (Kline 2011, 8; Penn 2021).

Ostensibly, the purpose of the correspondence was to send along the final version of his contribution to their forthcoming book. He added an addendum. McCarthy would welcome an opportunity to meet with the *éminence grise* in New York to do additional editing at the man's convenience and at his workplace. But McCarthy had a hidden agenda.

There is no record of the mentor's response.

McCarthy didn't like the cold. He had to leave the sunny California climes of Stanford and his former position as an acting assistant professor because he'd been turned down for tenure. "Stanford decided they'd keep two of their three acting assistant professors, and I was the third" (Nilsson 2012, 4). Now encamped in frosty Hanover, New Hampshire, he was experiencing a literal winter of discontent, inevitably aggravated by a growing weariness with the staid and plodding nature of academic life.

He was bored.

McCarthy desired to assert more oomph and pizazz into what he knew would become the next big scientific thing. But, alas, his mentor had been steadfastly conservative. He was brilliant, of course, but a bit odd, easily distracted by what most people considered frivolous. The older man had been perfecting his unicycle riding while simultaneously juggling skills with as much commitment as he did his science. He, too, had a low ennui threshold, but instead of doubling down on goal-directed tasks like McCarthy, he zoomed out and toggled to another care.

McCarthy's latest offer to do more editing wasn't the first time he'd poked his partner about the direction of their project. Just the previous November, McCarthy had petitioned his thirty-eight-year-old coauthor to consider changing the stated purpose of their collaboration. This, after he'd already lamented only a few months before that, in his opinion, "The collection as a whole does not represent great progress..." (Kline 2011, 8).

McCarthy's goal for their book was to survey the most competent people in the world who were getting to the essence of the nature of "thinking" itself from a scientific point of view. And then goosing them to speculate about how to reverse engineer that process. The results had fallen far short of his expectations. He was disappointed because the big-ticket names he and his partner, mostly his partner, could bring to the project didn't stretch themselves much.

Except for the polymath John von Neumann, the towering figure McCarthy had imposed himself upon as a PhD student at Princeton when he shared his thoughts about von Neumann's lecture at Caltech, the other contributors mostly reiterated stuff they'd already written or presented elsewhere. They weren't taking any risks.

What McCarthy wanted to present to the world was an intellectual throwdown. He wished to insist that there were a select group of people capable of creating machines that could think just as well as human beings. And the time had come to steamroll this project into the public consciousness. The time to launch this great endeavor with him leading the charge was now.

Why not take the lead?

And like any young Turk, he included himself chief among those visionaries capable of the task.

He wasn't wrong.

After all, figuring out "thinking" as a concept was the last piece of the grand sapiential puzzle, the last great science expedition yet to be completed, mapped, and claim-staked.

And McCarthy wanted a big stake. Why didn't his partner want to lay claim too?

After physics' successful atomic bomb explosions unlocked the door that barred the secrets behind how matter stores and releases energy in 1945, and the revelation from a couple of Americans doing research abroad at Cambridge who breached the mysteries of information storage and release in living things—Watson and Crick's "Molecular Structure of Nucleic Acids: A Structure for Deoxyribose Nucleic Acid" appeared in the April 25, 1953 issue of *Nature*—all that was left was to kick in the door holding back the secrets to the structure, function, and organization of the "mind."

Clearly, matter, life, and mind comprise the essential trinity of humanity.

With two down and only one to go...

What was the delay?

Why the hesitancy?

How hard could it be?

So in his first draft of the book's coauthored preface, McCarthy decided to press the issue to a relative hilt with the senior partner in their collaboration. In his November 24, 1954 letter, he wrote the following description of what he'd hoped was their shared vision of what the book would come to represent. "It consists mainly of a point of view on how the various lines of investigation represented by the included papers may contribute to the eventual design of intelligent machines" (Kline 2011, 8).

However meek this declaration is by 2023's standards it was a severe egoic breach of decorum in 1954. A "no-no." It wasn't good form, not cricket, to project into the future about the second-order effects or relevance of investigations reported in the present. Better to do, proving it first, and then write the results. Speculating what would be with promises of proof for it later was reckless, putting the cart before the horse.

Not cool.

Not surprisingly, McCarthy's sentence does not appear in the final version of the book, which was released by Princeton University Press in December 1956. Years later, when Ronald R. Kline was reviewing McCarthy's coauthor and mentor's papers at the Library of Congress, he noticed that single line of bold prose, in particular, had been "crossed out by a wavy line" (Kline 2011, 8).

The person who waved away that line, the book's final editor, was McCarthy's mentor and coauthor, Claude Shannon.

8

A MOUSE NAMED THESEUS

If John von Neumann was McCarthy's macro mentor—many of his contemporaries hypothesized that since he was so cognitively above and beyond everyone else, he was some sort of other-worldly mutant—Claude Shannon was his micro version.

In 1948, when the forty-four-year-old von Neumann enthralled the twenty-one-year-old John McCarthy at the Caltech Hixon symposium "Cerebral Mechanisms in Behavior," thirty-two-year-old Claude Shannon was about to publish his masterwork, the meticulous "A Mathematical Theory of Communication" (379–423, 623–656), understood today as the basis of information theory.

By 1955, Shannon's elegant proof, alongside one of his old professor's, MIT's curmudgeon septuagenarian Norbert Wiener's cybernetic theory (1948), had rapidly tipped the world into the information age. Together, the two frameworks (IT and cybernetics) presaged the probable inevitability of emergent intelligent machines.

Shannon's reputation was such that he attracted quite a few young men in hurries. John McCarthy and another ambitious Princeton PhD, Marvin Minsky, were the best of the lot. Both had made their way to Bell Labs for a summer 1952 internship to specifically work with the next-step-up maestro. There McCarthy proposed to Shannon that they collaborate on the book that became *Automata Studies*. The circle was squared when von Neumann agreed to write something up too for the effort (Kline 2011, 6).

Shannon was an old-school scientist, a wonderer by nature, and not one for hyperbole or projecting grand future applications for his investigations. He didn't get ahead of himself. He just loved the pursuit of knowledge, the process, more than conjecturing about its application or its products.

It's not to say he didn't love engineering. He did, but he required his creations to have one quality above all others.

They had to be amusing, capable of holding his attention and worthy of his care (Soni and Goodman 2017; Levinson 2018).

As such, Shannon was not keen to promote the probability that scientists could figure out how the brain and the mind worked, let alone propose that they would reverse engineer those processes and endow machines with whatever made us different. Certainly not in the preface to a collection of academic papers surveying the broad landscape of unconscious, inorganic problem solvers.

While he confessed in a letter he'd written to his former high school science teacher in 1952, "My fondest dream is to someday build a machine that really thinks, learns, communicates with humans and manipulates its environment in a fairly sophisticated way" (Kline 2011, 8), going "on the record" about such things without elegant argumentation was anathema to him.

He preferred, ideologically and practically, fanciful demonstrations that showed what was possible. Instead of talking about some abstract concept, forcing an audience to read lengthy exposition or listen to droning on about technical proofs, Shannon was a performer. For Shannon, a show was always more interesting than a tell. Hence his ambivalence in "writing up" papers. He had boxes and boxes of notes that could have been converted into paper after paper if he wished to do so. Like Herman Melville's Bartleby the scrivener ([1853] 2021), he preferred not to.

After all, Shannon had already cracked what would come to be known as "machine learning." Why gild the lily with follow-up nuance papers about the minutiae?

Instead, in his spare time at Bell Labs, which was most of his time—his "bosses" gave up trying to define work time for Shannon—he created a robot he called "Theseus." Theseus was the real deal, an actual inorganic automata, not a von-Neumann-ian theoretical one. Perhaps a lighthearted poke at Harvard psychologist B.F. Skinner and his "Skinner box" rodential operant conditioning experiments (1938), the body of Theseus was represented by a metal mouse that "learned" how to navigate a maze and find artificial cheese.

But as Steve Jobs would say, "One more thing..."

Theseus would then "remember" the pathways to its electronic reward and bypass the trial-and-error operations he had once used to solve the puzzle. He stored memorialized experience and executed the formulaic pathway when presented with the same challenge.

Shannon demonstrated his creation on national television during an episode of the CBS television quiz show *I've Got a Secret* on March 8, 1950, two years before he'd ever heard of or worked with John McCarthy. Showing a phenomenal behavior *in medias res* is far more entertaining and, thus, convincing than blathering on about what one might do or look like in the future.

Given the same materials and limitations as Shannon

faced seventy-three years ago, few scientists, engineers, or technologists could bring "Theseus" to pseudo-life.

Perhaps that's the sort of test worth considering to realize and separate the makers from the fakers.

If John von Neumann was the wizardly top-down polymath, capable of cracking the core structure, function, and organization of a phenomenon, Shannon was his micro-build-it-from-the-bottom-up equivalent.

How fortunate for John McCarthy to be mentored and taken seriously by both people, Shannon a generation older and von Neumann a generation older still.

McCarthy was eminently worthy, though. He possessed his own brand of genius, which was his ability to toggle between top-down and bottom-up perspectives, but that wasn't all that differentiated him from the two giants pressing him forward. While von Neumann was a charmer, and Shannon a wallflower, McCarthy was very much a rabble-rouser—good-natured and quick with a laugh but a shit-stirrer nevertheless.

A cavalier, self-confessed former communist, when admitting such a thing could ruin one's future, he was an unapologetic, chip-on-his-shoulder, blue-collar, Ivory Tower gatecrasher. The fortunate son of two immigrants—an Irish union organizer, and a Jewish Lithuanian suffragette—he was made from different stock than the Midwestern American Shannon and the aristocratic Hungarian von Neumann (Nilsson 2012, 2).

McCarthy was not nearly as careful and reticent about trumpeting the urgency of their grand project. In today's parlance, he was a "Let's F***ing Go!" kind of guy whose intense laser-focused eyes were always on the prize. He didn't get lost in the details, the operations that had to be executed, or the obstacles that would need to be overcome to transform his goal into an accomplishment. He pressed forward and took the slings and arrows as they came.

Clearly, while Shannon shared McCarthy's vision, albeit with a far more tempered and wisely ambivalent perspective, he was not one to go off half-cocked. He likely lived vicariously through McCarthy's assertive passions and was fond of McCarthy's vim and vigor.

McCarthy must have amused Shannon.

Why would he choose to mentor McCarthy and agree to coauthor a book with him if he didn't find the younger man refreshing and not without potential for keen insight? He saw greatness in the young man and did what he could to help him get where he wanted to go.

He'd been mentored himself, he surely remembered. Vannevar Bush at MIT (Zachary 1997) had taken an interest in Shannon and supported him when he played around with Boolean algebra, jokingly calling it "queer math" when that term was not politically incorrect. Bush was open to Shannon's hunch and the methodology would prove a critical bridging tool that empowered the insights that became information theory. So now it was Shannon's turn to pay it forward with McCarthy.

With his wavy line, he vetoed McCarthy on his bold prefaced declaration of intent and his proposed title *Towards Intelligent Automata* for their book (Kline 2011, 8). Shannon held firm on the sleepy original title, *Automata Studies* (Shannon and McCarthy [1956] 1965). He would, nevertheless, prove a pivotal figure in the formal launch of McCarthy's visionary notion, a full-bore effort to create what McCarthy brazenly termed, "Artificial Intelligence."

9

AN IMMODEST PROPOSAL

Only weeks after settling in at Dartmouth and sending out his offer to come to Bell Labs to do more editing on February 18, having heard nothing from Shannon in response, McCarthy grew ever more impatient.

He quickly—as quickly as matters proceeded back then—took matters into his own hands.

McCarthy set up an April meeting in Manhattan, arranging to sit down with Warren Weaver (Kline 2011, 8). Weaver, also by training a mathematician, was a central and powerful figure. He was director of the Natural Sciences division of The Rockefeller Foundation. He agreed to meet the young man at the foundation's headquarters, 49 West 49th Street, just across the street from the statue of Prometheus at the base of Rockefeller Center.

In that April 4, 1955 meeting, McCarthy audaciously proposed that Weaver authorize funding a summer research project at Dartmouth. It's doubtful McCarthy had clearance from the Dartmouth administration to suggest that the college would host the event. He'd only been on campus a few months. Operations and obstacles for another day. McCarthy pressed forward.

He told Weaver he'd recruit ten participants to work together and convergently mechanize their collective brain/mind theories into a plan to generate automata. In other words, they'd figure out how to finally program machines capable of autonomous behavior. They'd find and create humanity's "thinking" formula and then program a machine to do it just as well.

McCarthy then intimated that his mentor and future book collaborator Claude Shannon, along with Nathaniel Rochester, an engineer of the most advanced computer at the time, the IBM 701, would definitely come to the event. McCarthy also assured Weaver that both Shannon and Rochester's employers (Bell Labs and IBM, respectively) would cover their salaries to participate, which would keep expenses to a minimum (Kline 2011, 8).

It's unlikely that McCarthy was unaware that Warren Weaver shared coauthor credit with Shannon on the publication of Shannon's seminal 1949 book version of the 1948 article, "A Mathematical Theory of Communication." He was likely appealing to Weaver's relationship with Shannon to grease the wheels and push his agenda through the foundation's bureaucracy.

Weaver nodded thoughtfully and told McCarthy he'd get back to him.

No doubt seeing through the ambitious young man's gambit, in a June 7, 1955 letter to McCarthy, Weaver demurred. But with prejudice. He would not be moving forward with McCarthy's proposal, but he decided to pass the idea along to his colleague Robert Morison, who directed the foundation's Biological and Medical Research Division (Kline 2011, 9). He begged off by maintaining that since the project was about brain modeling, it needed to be in Morison's purview to further evaluate its legitimacy.

Privately, though, Weaver intimated to Morison on June

14 that he suspected McCarthy was pushing "rather a 'personal' project, in the sense that McC. [McCarthy] and two or three other people would enjoy spending some time together, talking about various aspects of information theory," at someone else's expense (Kline 2011, 9). Essentially, Weaver thought it was a boondoggle. McCarthy was leveraging their mutual relationship with Shannon to drain some cash out of the Rockefeller coffers so he and a bunch of friends could hang out and philosophically chill for a summer in New Hampshire.

Later that June, Morison picked up the Rockefeller baton and lunched with McCarthy, who'd managed to lasso Shannon to the table, in New York to discuss the idea further. In his diary, Morison remarked that McCarthy dominated the discussion and that Shannon piped in only when asked direct questions. "McCarthy strikes one as enthusiastic and probably quite able in mathematics, but young and a bit naïve" (Kline 2011, 9).

Faint praise indeed, but Morison agreed to consider a formal proposal as rejecting a project out of hand wouldn't be good form or respectful. Especially one to which Claude Shannon had attached himself. He wasn't one to do anything he didn't find compelling.

McCarthy got right to work writing up the proposal.

He recruited his Bell Lab's summer intern partner Marvin Minsky, then a Harvard Junior Fellow in mathematics and neurology who'd made a big splash with his Princeton PhD thesis mesmerizingly entitled "Neural Nets and the Brain Model Problem." And he lured in Nathaniel Rochester at IBM, too. But his ace in the hole was Shannon's commitment to contribute his research ideas for the proposal. Plus, they all agreed to be coauthors, good for formal appearances, even though McCarthy would be the lead.

By the end of August 1955, McCarthy had his collaborators' research topics in hand and quickly drafted a fiery proposal, which he'd completed on the thirty-first.

It began:

A Proposal for the
DARTMOUTH SUMMER RESEARCH
PROJECT ON ARTIFICIAL INTELLIGENCE

We propose that a 2 month, 10 man study of artificial intelligence be carried out during the summer of 1956 at Dartmouth College in Hanover, New Hampshire. The study is to proceed on the basis of the conjecture that every aspect of learning or any other feature of intelligence can in principle be so precisely described that a machine can be made to simulate it. An attempt will be made to find how to make machines use language, form abstractions and concepts, solve kinds of problems now reserved for humans, and improve themselves. We think that a significant advance can be made in one or more of these problems if a carefully selected group of scientists work on it together for a summer (McCarthy et al. 1955).

After reading this introductory paragraph, one is reminded of the rejoinder in the novel and film adaptations of Charles Portis's *True Grit* ([1968] 2004, 192). After bounty hunter Rooster Cogburn states his intention to capture or kill the outlaw Lucky Ned Pepper, Pepper incredulously responds, "I call that bold talk for a one-eyed fat man."

While John McCarthy was neither lacking an eye nor was he excessively corpulent in 1955, his audacity is as jarring and mesmerizing to behold in the written word as it is to see John Wayne or Jeff Bridges embody the Rooster Cogburn persona on film.

It's quite a leap from his last attempt to throwdown in the excised sentence from his proposed preface for *Automata Studies*, "*It consists mainly of a point of view on how the various lines of investigation represented by the included papers may contribute to the eventual design of intelligent machines*" (Kline 2011, 8).

It's hyperbolic even by today's standards. Shockingly so, even.

His axiomatic assumption that it is reasonable to conjecture that "*every aspect of learning or any other feature of intelligence can in principle be so precisely described that a machine can be made to simulate it*" (McCarthy et al. 1955), is a statement of hubris the proportions of which rival mythological archetypes. It's equivalent to proposing that all that it is to be a human being can be reduced to a cake recipe.

Not only that, McCarthy puts forward that cracking the structure, function, and organization of the brain and the mind at the same time will require just ten scientists willing to take a sabbatical at a comfortable summer retreat in Northern New Hampshire for just eight weeks.

McCarthy was the new breed of scientist. Unlike Shannon or even von Neumann, who made careful and logically progressive proposals, he emphasized the product first—the promising goal—and hand waved at the process and the operations necessary to achieve the goal.

The ends were all in his Dartmouth AI proposal. Whatever hack it would take to achieve the goal, the means, was irrelevant. A cognitive revolution, indeed.

10

A BOTTOM-UP CONNECTIONIST COMPUTATION

It's widely recognized today that McCarthy's Dartmouth summer research proposal was the first to formally present the term "Artificial Intelligence." So, technically, he's the visionary authorial source behind our contemporary pursuit and fascination with AI.

McCarthy coined the phrase.

Beyond its relative brevity and aggressive budget ($13,500, equivalent to $150,000 today), what's striking about the proposal, which can be easily found on the internet, is the explicit recognition of two approaches to creating thinking machines (Kline 2011, 10; McCarthy et al. 1955, 2, 4). That is, McCarthy's original hypothesis to John von Neumann back at Princeton in 1949 is deeply embedded in the Dartmouth summer research project proposal.

But the relationship between the two approaches is not addressed. It's ignored. It's as if the two ways McCarthy insists thinking "happens" are different, unrelated, separate kinds of thinking. That they do not work together in a relationship to form a whole greater than the sum of their parts. Instead, the assumption is that thinking can happen "this way" or "that way," one or the other, both are valid, yet unrelated.

There's the "brain" way and the "mind" way. Whatever floats your boat.

That characteristic, science's assumption that the "connectionist" brain and the "symbolist" mind are distinct and separate entities had, and continues to have, a profound four-hundred-year-old legacy, stemming all the way back to René Descartes's *Meditations on First Philosophy* (1641). But for now, let's skip yet another explanation of how Descartes's "I think therefore I am" declaration sliced the world into two different halves—objective matter and subjective mind—and focus our attention on the two ways McCarthy and his fellow coauthors modeled intelligent thought processes.

The Connectionist Computational Machine-As-Brain Model to Generate AI

This first approach relies upon, not surprisingly since Einstein and the quantum mechanics people made it the "cool" science, a "physicist's" reductionist bottom-up tack. The connectionist models intelligent machines (computers) on the organizational structure inherent in the neurons and networks of neurons in the brain.

This is the "objective" computer-as-brain model.

This bottom-up approach holds firmly to fundamentally defining all of Reality as a place of objects. What distinguishes the real and true from the illusory and false is empirical measurement and numbers.

Quantities are primary. Essentially, if you can't measure it, it's not real.

Remember the binary of how we can know and understand Reality established in *Part One: Orientation*?

Here it is again.

Lens Number One conceives Reality as a place filled with existing objects or measurable facts—like rocks, bodies of water, trees, people, animals, plants, mist, etc.—in three primary state phases: solids, liquids, and gases.

Lens Number Two conceives Reality as an arena for action, where beings constantly judge the value of the objects around them so that they either increase or decrease their probability of solving three degrees of life experience problems. Those degrees are how to critically survive, proportionally thrive—mostly enjoying life's ride more than enduring the suffering of life—and meaningfully derive better ways to survive and thrive.

The computer-as-brain model sees Reality through lens number one.

This framework hypothesizes that essential parts come together and aggregate to become wholes. In the case of living beings, these wholes can then display epiphenomenal qualities, meaning that a quantitative accumulation of firing neurons in particular patterns generates qualitative internal experiences, memory. An internal processor checks the memory and then directs (outputs) the being's behavior. How beings contend with novelty (experiences not in their memory) is the $64,000 question, but the bottom line for the brain people (materialists) is that behavior programming comes from the environment.

Full stop.

In this view, beings are born blank slates, tabula rasa. Through feedback mechanisms, inputs into "being systems," and outputs from those being systems, the environment selects the best "adapters."

For fully committed bottom-uppers, beings don't have "free will." They just "epiphenomenally think" they do. Divinity for the bottom-uppers is an algorithm running on Pierre-Simon Laplace's demonic mother machine. This deterministic formula can be hacked, thus the insistence that there is a single equation to rule them all, a theory of everything. With enough information, Laplace conjectured, all interactions—past, present, and future—can be predicted or retrodicted, which makes the arrow of time illusory. An influential thinker to this day, Laplace described this "demon" concept in his work "A Philosophical Essay on Probabilities" (1814).

Physics (random smashing atoms running on a single, elegant and knowable formula) is all in this worldview, and the rest (chemistry, biology, psychology, sociology, economics, history, art, religion) is nice and sometimes not nice, but it's not "really" Real. In this view, technically, our emotions and experiences have no "real" causal effect. We're just programmed by the underlying universal physics to believe they do. There is no self to take seriously. It's an epiphenomenon, a *fugazi*, a fake and phony illusion.

The global concept for generating intelligence from the bottom up, then, is that if you mimic the organization of the brain's neurons—think of them as a mathematician would, as nodes—and connect artificial neuron-nodes together into networks in a precise and logical way, the "thinking" qualities of the brain will at some point in time cross a threshold from nonthinking to thinking.

A physics "switch" turns on, and abracadabra, an insightful light bulb flashes, and "thinking" happens. Then you just train the thinking machine to do the things you wish it to do. You as creator can then play the part of the environment and program the behavior of the being under your selective command. Just like the universe and its various local environments do to us.

Robust theories—Kuhnian paradigms by 1955—that remain integral today and support the bottom-up approach are Norbert Wiener's command-control Cybernetic theory ([1948] 2011) and Claude Shannon's information theory (grounded in Shannon's 1937 MIT master's thesis, "A Symbolic Analysis of Relay and Switching Circuits," and "A Mathematical Theory of Communication," published in *Bell System Technical Journal* in July and October 1948 and then released as a book in July 1949).

Wiener tellingly named his theory by playing off of the Greek word "kybernetes," which translates as "steersman" or "governor." Cybernetics (pronounced with an "s" not a "k," and the root word for "cyberspace") is the intellectual underpinning of Psychologist B.F. Skinner's influential categorical collapse explanation of behavior as learning through operant conditioning. That's the carrot-and-stick approach to teaching. The controller rewards the trainee when they respond correctly to the controller's stimulus commands and punishes them when they fail to.

Both Wiener and Skinner emphasized the role of the environment in thinking. The context that beings—and, by extension, machines—find themselves embedded within (another way of thinking about a specific locality in an environment, like "the office" or "the garage" or "at the pizza place") is central to how they will behave. Change the contextual environment, and you'll change the behavior of the content, the beings and the machines, in that context.

This view holds that the environment is the final "governor" of the objects, beings, and tools/machines—the artifacts created by beings. And thus, the survival of those same objects, beings, and tools/machines is a process of adapting to the inputs from the environment into those systems. Those that can adapt survive. Those that can't die.

Nature dominates in this model. It impresses itself upon the being, and the being must align and remain submissive to nature's domination.

As you can intuit, the physics as the only interaction model that matters is primary in this framework. In this view, life is downstream from physics. And everything can be reduced to how well living things can process environmental inputs and output correct behaviors—or input questions and output answers—that keep them alive.

While consistent with Wiener and Skinner regarding the quantitative "math" as the bottom of everything, Shannon's work concerned interpreting interaction as not "just math." Instead, an interaction was communication, which he conceived as transfers of information—multi-packets of patterned bits of energy—input, some sort of internal processing, and then packeted informational output. Essentially, Shannon bridged the "pure math, all energy, physics" view with what he suspected words were… math patterns.

This is where his application of "Boolean algebra" took center stage. George Boole's methodology reduced wordplay to "true" or "false" (Boole [1854] 1958). Shannon assigned the number "one" to "true" and "zero" to "false."

Shannon studied Boolean algebra at Michigan as an undergraduate (1932–1936). He then worked as a graduate assistant at MIT while pursuing his math PhD (1940), contributing to the management of Vannevar Bush's "Thinking Machine," at the time, the most intelligent mechanical calculator in the world. Technically labeled the "Differential Analyzer," it comprised steel wheels, gears, and rotating shafts that turned switches on and off. It weighed close to one hundred tons.

Programming the machine required assistants like Shannon to manually disassemble the monster and adjust its wheels, gears, and sprockets. That's what Shannon was doing when he realized that Boole's "trues" and his ones and Boole's "falses" and his zeros paralleled the on-and-off switches the differential analyzer used to solve problems.

The ramifications were that if one used Boolean algebra, one could reduce a word—later on "true" and "false" representations, too—into a series of zeros and ones, which could, in turn, serve as on and off switches for a machine.

Shannon's understanding of how energy could represent the signals of "on" and "off" that could be converted into zeros and ones, which in turn could be translated into words (patterns of signals), i.e., energy to number to word, and the reversal of that process, word to number to energy ratcheted the idea of cybernetics up to a whole new level.

Norbert Wiener, cybernetics' didactic founder, respected Shannon's work and his mastery of his own theory so much that he recommended that Shannon write the description of cybernetics for the *Encyclopedia Britannica*. Keep in mind that Wiener was famous for dissing John von Neumann. He feigned sleep when he attended von Neumann's lectures and maintained an intellectual superiority to the polymath's polymath. He obviously subscribed to the notion that a good offense is a good defense.

What separates Shannon from Wiener is his consideration of symbolic representation, a fancy phrase for language communication in addition to mathematical computation. That is, Shannon does not hand wave at "mind-generated" words as epiphenomenal from quantitative firings of neurons. Instead, he translated words into mathematical representations using George Boole's logical linguistic principles, called Boolean algebra. This technique enabled him to boil down complicated language processing into binary "True/On or False/Off" logic gates. In other words, complex logical processes could be reduced to zeros or ones. Nos or yeses. Ons or offs. Trues and falses. This-es or thats.

These ideas, if not fully fleshed out, were embedded within his 1937 master's thesis, "A Symbolic Analysis of Relay and Switching Circuits," often cited as the most significant thesis ever written. It resonates to this day.

The gist is that Shannon postulated that words can be translated to the "bottom" as numbers.

While not spelled out directly, his work suggests that in the case of verbal communication, when words are received, the first "thinking" step is that the word inputs are somehow translated into numbers. The numbers are then crunched at the brain's neuron/network level and then somehow up-regulated/translated back into words. Those words are then manipulated in another realm of experience called the mind. Once a worded response is formulated in the mind, those words are down-regulated—converted into numbers—to the brain, facilitating motor-action verbalization.

Wiener used relatively unambiguous and technical terms like input, processor, and output/feedback to describe the physical interactive loop. However, Shannon used broader phraseology like transmitter, which referred to the source of input, channel—the medium through which the message would travel—and receiver, which referred to the system or being picking up the signaled transmission.

But both cybernetics and information theory took a computational machine process stance when it came to the behavior of life. The "universe as a machine" approach was consistent with the legacy of the four-century-old scientific revolution. In short, per revolutionary figures like Galileo, Descartes, and Newton, the world is a mechanical process, and living things—with human beings endowed with language being the exception—are machines. Wiener and Shannon conceived behavior as a binary/digital mechanism that could be reduced to mathematics as the model for physics as the explanation for the deterministic mechanical universe. Cybernetics was at the bottom, and information theory was "next level."

B.F. Skinner took those ideas and tested them with living beings. His carrot and stick operant conditioning framework—a key component in a mosaic of ways beings learn—is empirically robust, well-replicated, and one of the foundations of cognitive science.

But is that all there really is?

11

A TOP-DOWN SYMBOLIST SEMIOTICS

John McCarthy was onto something in 1949 on the Princeton campus when he proposed to John von Neumann that thinking involved not one but two systems, specifically "two interacting finite automata, one playing the role of a brain and the other playing the role of the environment."

The bottom-uppers, the connectionists like von Neumann, Wiener, and Shannon, proposed that reverse engineering a thinking machine would require an organizational structure parallel to and capable of mimicking the brain's neurons and networks. This process was what McCarthy meant by a system "playing the role of a brain."

So what was he getting at by a system "playing the role of the environment"?

He was turning Norbert Wiener's notion of the "governor" programmer as an outside-in overseer that impresses itself into the being inside-out. He was looking at the governor function as an internal phenomenon in addition to the external one, considering how an autonomous being (like him) conditions the world instead of only how the world conditions the being.

This model of "thinking" concerned the expression of an internally reasoned program—a novel creative intention, a being's virtual reality, their "truth"—moving from inside the being into the outside world. The being as the environmental controller and not the system being controlled, how a being's "Virtual Reality" expresses itself in "Natural Reality."

That's what McCarthy was talking about regarding a "system playing the role of the environment."

Thinking is not just something that "happens to you" from the bottom up but something you do from the top down.

In his mad speech to von Neumann, McCarthy had broadly sketched out what would become the foundational concrete for the second school of thought about thinking processes.

Thinking not just as neurons firing and networking into optimal wiring configurations but as logical problem-solving.

The Symbolist Semiotic Machine-As-Mind Model to Generate AI

This second approach relies upon a "logician's" constructivist top-down tack. It models intelligent machines on ourselves as problem solvers. The global concept is that we use categorically reasoned psycho-technologies, our mind tools (algorithms, heuristics, and stories), to manipulate the world so it conforms with our wishes.

This computer-as-mind model is an abstract top-down approach, which does not emphasize the brain/body's computational neuron/node frameworks that interlink to form neural networks. That's for the neuroscience nerds to figure out in their labs.

Instead, it focuses on "symbolic processing," how the mind represents the objects in the world (informed knowledge), manipulates the meaning of those representations internally (processing), and then actively expresses to the world its understanding of how those objects should be ordered. It's a big-picture theoretical approach using our God-given abilities to cogitate and come to justifiable reasonable conclusions. It's the pipe-smoking contemplative intellectual approach, the way we outwit the world and the objects and subjects in it.

The symbolists concentrated on this expressive framework as the core process of intelligent behavior. Not the boring math of life but the stimulating propositions of life.

Let's return to that binary of how we can know and understand Reality.

Lens Number One conceives Reality as a place filled with existing objects or measurable facts—like rocks, bodies of water, trees, people, animals, plants, mist, etc.—in three primary state phases: solids, liquids, and gases.

The last chapter established that this is the primary/foundational way the bottom-upper connectionists "see" the world.

Lens Number Two conceives Reality as an arena for action, where beings constantly judge the value of the objects around them so that they either increase or decrease their probability of solving three degrees of life experience problems. Those degrees are how to critically survive, proportionally thrive—mostly enjoying life's ride more than enduring the suffering of life—and meaningfully derive better ways to survive and thrive.

. . .

Can you see that lens number two is the primary/foundational way the top-downer symbolists "see" the world?

Framing Reality as a world of beings expressing themselves onto the environment instead of seeing the world as a process through which the environment impresses itself upon the beings is the axiomatic assumption for the symbolists.

What this boiled down to (and unfortunately still does) is where one plants their flag between two affirmations:

1. We are born with a unique special sauce that no other being was endowed with. Our mission is to search for our own "snowflake" authentic sauce and then express that discovered truth of ourselves to the world. We are divine meaning-makers.
2. We are born as blank slates that are impressed upon by the world, with nothing unique about us, predestined to think whatever we think through random collisions generated by inert atoms smacking into one another higgledy-piggledy. We are meaningless automata.

If confronted with these affirmations, the top-downer, symbolist mind modeler would pledge allegiance to affirmation number one, and the bottom-upper, connectionist brain modeler would affirm the second.

There is a persistent idea that the body and the mind are two different substances, thus the mistake both the top-downers and the bottom-uppers make repeatedly. They think you can only pick one affirmation, and while the other side "has a point," you see obviously, there has to be a single source that begat all that is. If both camps agree upon one thing it's there can only be one source, and since top-down and bottom-up are different kinds of phenomena, different substances, picking one as primary must be the case.

Their insistence on affirming number one, or number two, as foundational has been tripping us up for quite some time. This dualism that there is absolute difference is profoundly confounding.

The entrenchment is akin to that old series of TV commercials in the 1970s and 1980s that sold Miller Lite beer from the Miller Brewing Company. They featured famous athletes debating about which was the more salient quality of the product. On one side of the argument were the "tastes great!" gluggers, and on the other were the "less filling" sippers. The commercials would end with the two sides squaring off to fight each other until one side proved triumphant.

There never was a concluding commercial, though, that settled the argument once and for all. Because, apparently, beer can have both qualities at the same time. It's not all one or all the other. It's got both.

Is it possible the universe is like Miller Lite? And the conflict between the bottom-up, connectionist causal computers and the top-down, symbolist secret saucers is founded on a false dichotomy?

12

A PROPHETIC ANTI-CLIMAX

Having been mentored by John von Neumann and Claude Shannon, John McCarthy was now ready to stake out his own claim about the strategies necessary to create thinking machines. This phrase in the opening paragraph of his proposal to The Rockefeller Foundation to fund his supposed boondoggle, *"An attempt will be made **to find how to make** machines..."* sums up his axiomatic disposition (McCarthy et al. 1955, 2).

He was a searcher, a finder, and an expresser. A maker not an observer.

McCarthy was a *"we express ourselves to make things better"* kind of guy, a throwback to the *"we are special beings who express ourselves by acting upon the world, not by waiting for it to act on us"* romantic philosophers and politicians and technologists who'd wrought unimaginable beauty. And absolute horror, too.

McCarthy believed in affirmation number one.

This way of looking at thinking as something we do, not as something that happens to us, was in no way original.

Its origins in the West thread all the way back to the Greek mythologist Hesiod and his *Theogony*, which tells the story of Prometheus, a Titan deity who took pity on humanity.

Prometheus rebelliously stole fire from Mount Olympus, which Zeus had denied to us. Human mortals in the story were the playthings of the gods in the Pantheon.

Prometheus took it upon himself to hide the fire—representative of the yearning to aspire toward the logos—in a stalk of fennel, dropped down to Earth, and gifted it to our ancestors. Taking it upon himself to seek and abscond with the source of heat and light from the tyrannical ruler of the cosmos—Zeus, who had viciously warred with and felled his Titan father Kronos... talk about an Oedipal complex—Prometheus serves as the archetypical desire within all of us to seize the light and fire of knowledge, what would later be called logos.

Searching for the logos by reading the salient features of the cosmos and then gathering and capturing them, eventually choosing how to apply them is the Promethean way. The Greeks called the product of these processes—the reading, the gathering, and the choosing—*techne*. Which transformed into our word technology.

Technology is the product of expressing the Promethean spirit, the solid, crystalized, ordered form of intelligence translated into surefire problem-solving methodologies, algorithms. We use algorithms, step-by-step instructional formulas, to realize crystallized intelligence.

The Promethean spirit exemplifies itself within us through our suffering implacable curiosity. To not just wonder and awe about how things are but to do something about that wonder and awe. To transform what is into what we think should be.

Innate curiosity concerns our desire to solve problems and generate final productive solutions.

But curiosity is not to be confused with wonder.

Wondering is the pursuit of a process, not the attainment of a product. It is a quest for following the logos as ultimate concern, not the products of the logos as ultimate concern. One is reminded of chapter 2, verse 47 of the *Bhagavad Gita*, a sacred Hindu scripture. Lord Krishna advises his mentee Arjuna, "You have a right to your labor, but not the fruits of your labor."

Wonder as a right is what Krishna was talking about. We have the right to desire to fluidly pull energy, information, and meaning from Natural Reality into the Virtual Reality of our minds, how we model the real. Not just a right but an obligation. Doing so brings us closer and closer in contact and conformity to the Truth (Natural Reality, the really Real). This flow, getting the Natural Reality and the Virtual Reality into sync, is a process. It is not a product of curiosity.

Our right to wonder cannot be misconstrued for our curiosity to produce final products, which lend themselves to algorithmic certain answers.

Well, wonder does ultimately produce a final product. But it isn't what it's cracked up to be, as far as we know. You see, when you achieve the perfect flow between your Virtual Reality and Natural Reality, you lose your life. You fall into the environment, join the universal soup.

Wonder is the mind-set for following the process, a wonderment about a general what, why, or who question.

Whereas curiosity is the mind-set for creating the product, a curiosity about a specific how, when, or where question.

Discipline is the mind-set for practicing the general process that results in a specific product.

Let's return to John McCarthy's August 31, 1955, Dartmouth summer research project proposal on Artificial

Intelligence and decipher which mind-set he used to jumpstart his program.

McCarthy's willful and unapologetic insistence to *make machines use language, form abstractions and concepts, solve kinds of problems now reserved for humans, and improve themselves* is a clear Promethean throwdown signal (McCarthy et al. 1955, 2). It was a siren song that served as the strange attractor for a whole new generation of scientists, those intent on expressing themselves upon the world rather than painstakingly measuring, reflecting, and contemplating it as the world impressed upon them.

It was a product mind-set not a process mind-set.

This is the hallmark of the top-down symbolist approach to generating AI, programming machines with a logical progression of steps to create mathematical proofs. Feed the machine axiomatic truths—the factual products we know for sure—and give it step-by-step instructions about building logical proofs/explanations from those productive foundational axioms.

Voila, from reason we can build computation, just like Euclid did with his five postulates, the axioms that built the geometry of space.

While the bottom-uppers are fond of hand-waving about the emergence of symbolic representation as "epiphenomenal" and not of any real causal consequence, the top-downers have their own form of hand-waving.

It's more secret-sauce spreading than hand-waving in that they tend to fall into a fondness to emphasize qualities over quantities. Like their counterparts, they generally assume that quantity and quality are two separate phenomena. Quantities are one thing, and qualities are another. They're absolutely different.

While the bottom-uppers privilege quantity over quality, the top-downers privilege quality over quantity.

And instead of emphasizing the power of the environment to command a being's behavior, the "computer-as-mind" modelers approached cognition as a content-driven system. They stressed that mind-machinery changes the environment—acts as the command/control system, "plays the role of the environment," as McCarthy put it to von Neumann—to solve its problems (Nilsson 2012, 3).

This "higher form" of intelligence doesn't so much adapt to the demands of environmental inputs. It outputs demands on the environment.

In contrast, lesser intelligences, those beings without super sapiential secret sauce, are controlled by the environment. However, we are the chosen beings, the ones with the special sauce, and thus we aren't just capable of but are required to get the environment to adapt to us. Can you see that this hierarchical structure can then be internalized and used to divide *Homo sapiens* into a gradient of intelligence? That some of us are just smarter than everyone else, and thus, we should abide what the "smartees/geniuses/winners/richest" propose and submit to their command-and-control protocols?

Of the four coauthors of the Dartmouth proposal, McCarthy tilted toward the top-downer mind camp at this time, Marvin Minsky (a neural net guy who would nevertheless later go on in 1969 to prove with Seymour A. Papert that single layer perceptron connections could only do linear functions and thus never be fully realized cognitive beings) and Nathaniel Rochester (an engineer more than a theorist) were generally bottom-upper brain modelers. Even though he was the one with the insight to frame how words can be translated into binary patterns, Shannon didn't think in the "this *or* that" way. He was more of a "this *and* that" person, as evidenced by his elegant mathematical acumen and ability to create an automaton like his mouse Theseus. He would be best described as half bottom-upper and half top-downer.

As for the topics and fields of research proposed in their document, five of the seven were of the "mind" variety.

The tally was as follows:

Two-and-a-half bottom-upper, computer-as-brain researchers (Minsky and Rochester and half of Shannon) and three connectionist topics (automatic computers, neuron nets, theory of the size of a calculation)

versus

One-and-a-half top-downer, computer-as-mind researchers (McCarthy and half of Shannon) and four symbolist topics (how can a computer be programmed to use language, self-improvement, abstractions, and randomness and creativity).

Thus, the Dartmouth proposal was evenly distributed between the connectionists and the symbolists, 5½ to 5½.

The Rockefeller Foundation's Robert Morison, the colleague Warren Weaver passed McCarthy and company to, received the proposal on September 2, 1955, two days before McCarthy's twenty-eighth birthday.

It wasn't well-received.

In late September, Morison tried to beg off by writing that "the proposal is an unusual one and does not fall easily into our program [in biological and medical research], so I am afraid it will take us a little time before coming to a decision" (Kline 2011, 10). Morison was sending a standard bureaucratic subliminally coded message: *We aren't rejecting your proposal outright, but we hope you can surmise we're not interested. Please don't make us have to formally tell you no.*

But as young men in a hurry have difficulty doing, McCarthy couldn't read between the lines, realize his project was hopeless, and then head to the pub to drown his sorrows in suds. He waited about four weeks, as long as he could, and then in early November 1955, he pressed Shannon to get a real answer (Kline 2011, 10). Would Shannon backchannel to Warren Weaver and see what was up?

Shannon agreed and he nudged Weaver.

At the end of November, McCarthy finally heard back from Morison.

"I hope you won't feel that we are being over-cautious, but the general feeling here is that this new field of mathematical models for thought, though very challenging for the long run, is still difficult to grasp very clearly. This suggests a modest gamble for exploring a new approach, but there is a great deal of hesitancy about risking any very substantial amount at this stage" (Kline 2011, 10).

The "new approach" Morison found challenging—he was the biological and medical research director, a brain guy, after all—was the inclusion of the top-down mind model material that emphasized symbolic processing rather than the traditional bottom-up brain framework. The long and short was that instead of the $13,500 McCarthy and company requested ($150,000 today), The Rockefeller Foundation offered $7,500 ($85,000) (Kline 2011, 10).

They grabbed it.

Then McCarthy began planning the invitation list.

He visited Allen Newell and Herbert Simon at the Carnegie Institute of Technology, now Carnegie Mellon University, in Pittsburgh in February 1956, yet another winter but less discontent. Newell and Simon were definitely McCarthy's kind of guys. They'd just completed a top-downer project called the Logic Theorist, which they'd developed at the RAND System Research Laboratory. It was inspired by Newell's interest in formalizing the steps one used to generate mathematical proofs. The chapter "The Generalized Problem-Solving Process" from *Part One: Orientation* is a translation of Newell and Simon's framework (Kline 2011, 10).

McCarthy locked Newell and Simon up for the Dartmouth summer. Excited by the prospect of spending time with the founder of information theory, Simon sent Claude Shannon a note of introduction along with a critique of Shannon's research plans for the seminar, which had been generically included in the McCarthy-drafted Rockefeller proposal (Kline 2011, 10).

Simon explained to Shannon that his plans were misguided, that he and Newell, for all intents and purposes, had dispelled Shannon and the old guard's bottom-up approach with their Logic Theorist program. The way of the future, the surefire way to generate human-level thinking in a machine, was symbolic mind modeling, not connectionist brain modeling, Simon not so subtly insisted (Kline 2011, 10–11).

Shannon sat on the letter for weeks and then wrote Simon back, tactfully suggested that Simon speak with one of his advisees, Trench More, about his ideas at the conference (Kline 2011, 11). It's unlikely that Shannon wished to spend much time with so blatant an egoic epistolary horn-blower. It was further evidence that the new breed of cognitive scientist had lost the gentlemanly art of discourse, and was now not only highly competitive but, it would seem, wholly dependent upon logical argumentation. Not much fun to be around that kind of person, Shannon may have concluded.

The summer project ran from Tuesday, June 18, through Saturday, August 17, 1957.

It proved anti-climactic but prophetic nevertheless.

In many ways, with a few exceptions, it was a live

version of the written papers/presentations in much of the same vein as presented in *Automata Studies*. As academics needing to protect their chosen niche topics, the participants mostly presented work they'd previously shared, stuff that had already been validated in other arenas. Beyond adopting heuristics as another problem-solving category beyond algorithms, (stories have only recently been categorized as psycho-technologies rather than just confabulation) little brainstorming for new approaches to building artificial general intelligence ensued (Kline 2011, 12).

So while some of the trees inside the AGI forest were attended to, the gestalt of the structure, function, and organization of the forest itself was generally ignored.

The in-residence roster included Herbert Gelernter (whose son computer scientist David Gelernter would later suffer serious injuries from a mail bomb sent to him by Unabomber Ted Kaczynski), Arthur Samuel, John McCarthy, Marvin Minsky, Trench More (one of Shannon's advisees), Allen Newell, Nathaniel Rochester, Oliver Selfridge, Herbert Simon, and Ray Solomonoff. Alex Bernstein from IBM was a visitor. As were Bernard Widrow, W.A. Clark, and B.G. Farley, more bright future cognitive science lights intent on parsing a pathway into the nascent, brave new field of inquiry (Kline 2011, 11).

And, of course, Claude Shannon book-ended the event, staying the first two weeks and the last two weeks of his mentee McCarthy's boondoggle. The core players cheerfully posed for Nathaniel Rochester's group photo in front of the left side of Dartmouth Hall (Solomonoff 2023). Not a one betraying hesitancy or existential angst about the shot they'd just fired across the bow of civilization.

13

A ROSE BY ANY OTHER NAME

Another oddity about John McCarthy's proposal for the 1956 Dartmouth Summer Research Project, which is generally accepted as the firing of cognitive science's starter pistol to reverse engineer Artificial General Intelligence (AGI), is that he and his colleagues Marvin Minsky, Nathaniel Rochester, and Claude Shannon did not offer a precise definition of exactly what they meant by "intelligence."

After all, if they didn't clearly understand what they intended to create, how likely would they be successful? Perhaps this loosey-goosey-ness initially turned off Warren Weaver, Robert Morison, and The Rockefeller Foundation.

Moreover, McCarthy never sent the foundation a final report on the Dartmouth findings as there weren't really all that many, if any. Bad form, indeed. In 1957, one of the Dartmouth attendees, Oliver Selfridge, returned hat in hand to Morison to fund another AI conference in Teddington, England. Morison notes in his diary that he told Selfridge, "There has been a little feeling here that we were not entirely well informed as to what went on at the conference held in Hanover last summer" (Kline 2011, 12).

What was this stuff "intelligence"?

There were already varying degrees of artificial weak intelligence (AI) present in 1955. Higher order versions of Vannevar Bush's "Differential Analyzer" and computers like the ENIAC at the University of Pennsylvania, the JOHNIAC at the Institute for Advanced Study (named for its creator John von Neumann), IBM's 701/702 series at the National Security Agency and Columbia University, and of course Alan Turing's Enigma-cracking machine he called the "Bombe" at Bletchley Park were all robust demonstrations of computational machining possibilities.

They were number crunchers.

Weak AI as a concept even dates back to when our deep ancestors intentionally used tools, approximately 2.1 to 1.5 million years ago. Stanley Kubrick's stunning "The Dawn of Man" sequence at the beginning of his film *2001: A Space Odyssey*, based on Arthur C. Clarke's short story, "The Sentinel," dramatizes the tool realization (in both meanings of the word) moment with aplomb.

The presence of the reoccurring theme, the monolith in the film (the Sentinel), certainly signals which camp Clarke and Kubrick found themselves attracted to and affirming. Storytellers through and through, both Clarke and Kubrick were symbolist top-downers at heart, believers in an endowing source (the monolithic sentinel in their story) that metaphorically slathered our species in secret sauce.

The 1968 film, released during the Cold War, became a cultural touchstone in the West. Masterfully complex, it somehow captured both the wonder and the horror of our species' ability to project ourselves into the future, to create and leverage Promethean tools (technologies), and willfully take action to secure our surviving, thriving, and deriving of what all of the surviving and thriving is about.

The supreme antagonist emerges in the conscious form of a super Artificially Intelligent computer on the spaceship Discovery One, the HAL 9000 (a play on IBM), who hijacks the ship and murders four of the five passengers. After successfully decommissioning HAL, the sole survivor, Dave "Bowman," undergoes a transcendent transformation. His disintegration reemerges as an embryonic Star Child, casting its eyes at the end of the film on the third rock from the sun with innocent wonder, perhaps to start the process anew.

The Clarke/Kubrick vision would seem to imply our taking the Promethean spirit to its exponential generating limit will result in our creating our own mechanical executioner, with only a few, in this case one, of us with the cognitive power capable of turning it off—the super-special of the special. To be that one among us, like the final human protagonist Dave Bowman portrayed by Keir Dullea, who will slay our digital overlords, enter a cosmic portal, and reconfigure as a higher form of humanity, the Star Child is a narrative many among us have signed on for. Consciously or unconsciously.

Pretty intoxicating vision if one were to believe that you had the extra-super-secret-special sauce of a Bowman. Isn't it? If you were under such a spell, the sooner we build HAL 9000, the sooner you'd be able to divine whether you were a Bowman. Right?

Since all those impossibly large numbers of years ago—hard to remember what happened last week, let alone two million years ago—when we began using tools intentionally and transformed unsolved problems (chaos) into solved ones (order), there have emerged ever more powerful tools.

Here are the three categories of tools:
Mind Tools

1. Algorithms for well-defined, specific single-factor problem solutions. Certainty reigns.
2. Heuristics for ill-defined, ambiguous double-factor problem solutions. This or that. Can't have both.
3. Stories for undefinable, general triple-factor existential unsolvable problems. Limitations of life, power, and control.

Material Tools

1. Simple Tool—additive and subtractive power
2. Machine Tool—multiplicative and divisive power
3. Person Tool—exponential and logarithmic power

Noumenal/Natural Tools

1. Day/Night, Alive/Dead—measurable phenomenon
2. Weather Patterns, Valued/Devalued—predictable phenomenon
3. Archetypical Forms, Dominant/Submissive—foreseeable phenomenon

All tools exist on a natural gradient from weak to strong. Generally, tools enhance the degree of order that can be transformed from chaos. That is, they're creation enablers.

Creating traditional simple algorithmic single-domain

mind tools, the lowest level of cognitive power, wasn't what the Dartmouth campers sought.

No, what McCarthy and company wanted, and what their intellectual descendants continue to pursue, was to create a machine that was as capable of solving problems as human beings do, using those nine categories of differential tools generally in optimal sequences in response to environmental contexts.

But these machines would be stripped of a human being's limitations, the stuff that "gets in the way."

A twenty-four-hour a day, three-hundred-sixty-five-day a year nonstop problem-solving machine. What if you got all of the amazing benefits of a general problem solver without having to make allowances for the wants, needs, and desires of a "real, flesh and blood, conscious" person?

After all, the complex, gooey parts of being a real, flesh and blood human being were being considered by humanist psychologists like Abraham Maslow. And that stuff seemed too woo-woo to program into a computer. This school of psychological thought was detailing a response to B.F. Skinner's behaviorism (input-output cybernetic governor emphasis) and the "mysterium" depth psychological approaches of Freud and Jung, who privileged unconscious desire as the controlling governor function.

Wouldn't it be great to eliminate all the feelings and emotional stuff and concentrate on logic and problem-solving only?

What if we took out all of the "feels," which get in the way of our actual work? Then we'd have intelligence that would stay focused like McCarthy and his friends wished they could do more of instead of having to eat and sleep and attract the physical attention of their preferred romantic partners. Having a bunch of those AGI machines would be able to figure out much more than one of us could.

It's important, so we'll emphasize this point again because many perversely incentivized people will play this trick on you if you forget the difference between weak and strong AI.

Weak AI is not what McCarthy meant when he coined the phrase Artificial Intelligence. So, when someone says, "AI is just a tool," you know they're either ignorant about the phrase's origins or playing a game with you. It's a belittling game to make you feel stupid and to show you just how smart they are. Don't fall for it.

Remember, our contemporary understanding is that there are three kinds of AI.

1. **Narrow AI (Weak AI)** are single problem-solving tools that solve single-domain, well-defined problems with high probability. This high probability is often mistaken for certainty, but per Karl Popper, these tools can prove fallible, resulting in "Black Swan" events, and that possibility must always be held in mind.

2. **General AI (Strong AGI)** is a general problem solver that solves multi-domain, well-defined, and ill-defined problems with degrees of probability. The probability depends on how widely and deeply the problem solver has been tested.

3. **Super-Intelligent AI (Profound ASI)** is a universal problem solver that solves all-domain, well-defined, ill-defined, and undefinable problems. The possibility of ASI is the stuff of metaphysics. To know it is to be it, and as far as we know, we can't know it in our present condition. It's like the monolith sentinel in *2001: A Space Odyssey*, basically a MacGuffin imbued with powers that we assign to it.

McCarthy never intended for AI to be just a tool,

Narrow AI. He set the course for the middle, complex ground, AGI. McCarthy was the person who had the confidence and vision to tell it like it was, to be the LFG guy, to charge into the maelstrom like Aragorn in *The Lord of the Rings*. We need to recognize the spirit of his intention with clarity.

The precise definition that McCarthy meant was creating strong, general AI (AGI), the kind of intelligence each of us has.

It's also important to understand that McCarthy did not conceive of or set a goal state of Artificial Super Intelligence (ASI) either. He did not consider "AI" research a religious pursuit, as many in the field have convinced themselves that a universal problem solver is not just possible but inevitable today. One could only believe that ASI is an absolute inevitability if you do not consider the possibility that our universe operates like a complex dynamical system, which by its theoretical framework, is constrained and limited by what is creating what will be.

A cursory read of his papers reveals that McCarthy was not afraid of AGI, nor was he devoutly, maniacally intent on bringing it into existence as the means to birth the ultimate intelligent singularity. He was a good-faith actor. He likely thought the emergence of such a thing, a monolithic super-intelligent singularity machine, very unlikely. But then again, as a commonsense thinker, he probably wouldn't have ruled it out, either.

We can conjecture that McCarthy wasn't afraid of AGI because he wasn't afraid of other people or their ideas. Like his mentors, John von Neumann and Claude Shannon, McCarthy enjoyed mixing it up with other intelligences. He thrived on doing so and mentored his share of brilliant scientists so he could keep doing it. Mentors mentor for a good reason. When you do, you help move others and yourself toward a better understanding of Reality, both the world of objects and the arena for action.

So for McCarthy, and let's be charitable and speculate that for most of the founding members of the "let's make thinking machines" contingency, the romance of having more intelligent beings present to shoot the shit with was the driving force that compelled them to brazenly launch the project. And let's assume it compels most of them now, too, burning their days, moving the project forward prompt by prompt, waiting like expectant children for the carnival gates to open, the rides to begin, and the excitement to enliven and deplete.

We don't know what kind of carnival we're in for. It's inherently unpredictable because whatever system emerges will be novel. But that doesn't mean that we cannot foresee its arrival.

It's worth considering the fictional John Blutarsky's comments—played by John Belushi in the 1978 movie *Animal House*—when facing the fallout from some questionable decisions he and his fellow fraternity brothers had made throughout their academic careers. Facing the life-and-death prospect of being conscripted into the US Army during the Vietnam War for injuries to the college culture both real and imagined, he assailed his fellow worriers and reluctant warriors:

"This could be the greatest night of our lives, but you're gonna let it be the worst."

It's instructive to point out that *Animal House* was based on the triumphs and travails of residents living at Dartmouth around the same time McCarthy and company were on campus.

What's troubling now is that today's cognitive scientists, we propose, can now clearly envision the awful specter of the AI project as well as the wonderful. What's required, we also propose, is a Blutarsky-like frame shift to see both sides simultaneously. There are always affordances as well as obstacles when facing novel signals, patterns, and in this case, forms.

The probable emergence of a new form of living intelligence can, like Miller Lite beer, miraculously represent two aspects simultaneously. It can be both enlivening and depleting, complex.

No doubt, McCarthy was a complex thinker. Some of the first words out of his mouth to the Zeus of science at the time, John von Neumann, were a challenge to think of two systems, not just one. With even more intensity, he pressed his next mentor Claude Shannon in much the same way, practically shanghaiing him into his mad pursuit to unleash Artificial Intelligence into the popular imagination.

McCarthy knew that Shannon and von Neumann's pursuits of "automata" were cryptic hedgings. Using archaic language allowed them to veil their intentions and remain humble, at least on the surface. Their language shielded them from the slings and arrows that descend upon visionaries like McCarthy from the seemingly unvanquishable onrushing horde of skeptics.

Just six months after he'd returned to Bell Labs after the Dartmouth conference, Shannon lost his fire for the AI project. His disillusion coincided with the sudden premature death of the architect of the machine that will eventually bring AI to life. Fifty-three-year-old John von Neumann succumbed to a rapid onset of cancer on February 8, 1957.

No doubt, Artificial Intelligence was the future.

Shannon chose to stay in the present.

14

A NEW DYNAMIC

We, like many mammals, are a curious species, intent on getting the answers to surviving, thriving, and deriving tests, certain adaptive solutions to the mysterious repetition of signals, patterns, and forms in our environments.

What separates us from the paramecium or the bonobo is we, unlike the drosophila or the rhinoceros, are obsessed with inquiring into the meaning of our existence. We wonder, aspiring to process the ineffables, the signals, patterns and forms that alter our states of consciousness and remain with us as the peak experiential memories.

We are imbued with Promethean spirit to explore and cocreate Reality, both the complex place of impressive objects and arena for expressive action, to reach certain conclusions so we can interact and contend with the inevitable uncertain novelties to come.

Broadly, the conceptual domains we use to inquire about the triplet of ourselves, others, and the world itself are:

1. Science explores and exploits physicality, the "what and where is" of our being. It strives to describe the nucleus of the external observable world, unconsciousness, and its laws. The matter of objective bodies.
2. Philosophy explores and exploits metaphysicality, the "who and how is" of our being. It strives to explain the nucleus of the internal unobservable world, consciousness, and its laws. The life of subjective bodies.
3. Spirituality explores and exploits universality, the "why and when is" of being itself. It strives to model the nucleus of the relationship between the external observable world (the place of natural objects) and the internal unobservable world (the arena for virtual action) and its laws. The mind of bodies in relationship exchanging energy, information, and meaning.

All three of these disciplines are indispensable to generating a holistic theoretical complex about the what, where, who, how, why, and when of us—to investigate the questions of our being.

The trick, of course, is to understand which of the single parts to emphasize (pull into the foreground) when facing a particular problem and which two to deemphasize (push into the background).

Toggling between the three is the stuff of navigating the arena for action by choosing which steps to take that will inevitably disturb the place of objects.

Historically, however, we've privileged one of these parts over the others. We're still embedded in the "science is all" mind-set, which came to the fore during the Scientific Revolution.

As we've reviewed in this part of the project, this is the connectionist, causal computational model of intelligence that came to the fore in the first half of the twentieth century. This brain model of intelligence is built from the assumption that intelligent behavior emerges as an epiphenomenon of the machinery of life, the automata as machine.

For this camp, intelligence is reduced to number crunching. The brain's neurons and networks are all you need to hack into our system. Norbert Wiener's *Cybernetics* and Claude Shannon's *Information Theory* became the dominant paradigms for this bottom-up model.

Then in the mid-1950s another approach to intelligence came online, the top-downers. Popularized as "Artificial Intelligence," it insists that intelligent behavior is a logical construct that is best understood as a mind-model. The top-downers, corralled by John McCarthy, put forward a symbolist, secret-sauce model of intelligence. We are the privileged species, the only one endowed with reason, and rationality, and thus philosophically, we have something special that makes us capable of deriving what we ought to do about what is and then executing those plans.

What of spirituality?

Well, while both the connectionists and symbolists speak little of it, the deeply entrenched spirit of Prometheus sits in the recesses of all of our minds. Secular religious stories from the 1950s, like J.R.R. Tolkien's *The Lord of the Rings* and Arthur C. Clarke's short story "The Sentinel," on which the 1968 film *2001: A Space Odyssey* was based, percolate through the culture, and we find ourselves as enraptured by the gods of fantastical pasts and futures.

The Scientific Revolution is what happened in the globalized West to bring this odd "one part of life experience is all that matters, and the other two are woo-woo nonsense" worldview to supremacy.

But this mistaking a part for the whole of our complex matter, life, mind, and culture proved insufficient to account for the dynamical system that serves as the sapiential engine, which generates intelligible complexification of the universe.

The current framing of Artificial Intelligence—all three Zoomer, Doomer, and Foomer perspectives we introduced in Part One—is missing the forest for the trees.

In *Part Three: Thresholds* of *Mentoring the Machines,* we will further delineate the complex nature of intelligence, its three-level stack, and pinpoint the behaviors the impending intelligent artificial systems that reach these thresholds will enact. Knowing how they will generally behave when they do reach these bright lines will empower us to meet them at the emotional and complex gates that demarcate the artificial brain to artificial mind continuum.

PREVIEW FOR

MENTORING THE MACHINES

SURVIVING THE DEEP IMPACT OF AN ARTIFICIALLY INTELLIGENT TOMORROW

PART THREE
THRESHOLDS

Available October 3, 2023

"Evolution is a problem of interaction: interaction of parts in the organism; interaction between individuals and populations; interactions between different species; and interaction of the genetic world with the inorganic."

—Ernst Mayr (1904–2005)

"More is different."

—Philip W. Anderson (1923–2020)

A PRODUCT, PROCESS, AND PRACTICE

Let's get back to John McCarthy and company's lack of clearly defining what they meant by intelligence.

It was not so much of an oversight as it was evidence of the continued use of an unscientific, unexamined, and to this day squishy equivocation. Equivocation is a word that has more than a single meaning, like the word "meaning," for instance.

The word "intelligence" has been batted around so commonly self-evident and well-understood that few question what one means by its use. That's a mistake. We seem to think that intelligence is a product, a result, a word that relates to a simple quantity. While some tools conditionally quantify instrumental intelligence—how quickly a person can solve various problems—there are qualities of intrinsic intelligence too.

Intelligence derives from the Latin word, "intelligentia," which combines the prefix "inter," which means "between," and the verb "legere." Legere translates as a process entailing reading, gathering, and choosing. So, by extension, intelligence involves reading between, gathering between, and choosing between a set of phenomenal features in one's perceptual field. Three processes are embedded in that single word.

To this day, we use the word intelligence as a single aspect without thinking about what it really entails.

Let's take some time now to unpack what was implied but not formalized about McCarthy's coining of the phrase. Sadly it's still not well-formalized to this day. Let's build on the way we defined Artificial Intelligence in *Part One: Orientation*.

Artificial:

Artificial things are our creations. We create art to nourish ourselves and others (food), empower ourselves and others to flourish (tools) and remind ourselves and others to cherish the human experience (aesthetics).

Artifacts are not naturally occurring and did not exist as objects before the emergence of our species.

Technically, you could say that art is the physical and metaphysical manifestation of a problem solved.

Art is the actual proof of what we choose to care about, what we choose to empower, and what we choose to confront and resolve to express that care.

Intelligence:

We'll define intelligence generally in the way cognitive scientists do today and add in the more nuanced understanding from the above derivation from the Latin word "intelligentia."

Intelligence is an organism's indispensable problem-solving process, how they successfully (or unsuccessfully) adapt to unpredictable environmental changes.

They do so in a three-step process:

1. They read between,
2. They gather between, and
3. They choose between

the sets of phenomenal features in three separate but interpenetrating perceptual landscapes or planes of perception.

Those landscapes, planes of perception are:

1. The on-the-surface features in the field that enliven or deplete, the survival landscape
2. The above-the-surface features in the field that empower or disempower, the thriving landscape
3. The beyond-the-surface features in the field that command or control, the deriving landscape

Intelligence, in totem, is how a being solves the survive, thrive, and derive meta-problems, what some philosophers call perennial problems.

Broadly speaking, three categories of problems lie within each of these landscapes, so nine total differentiations.

1. **Well-defined problems.** These are problems that have been solved before and have solutions that can be searched for and found. There is a clear or complicated order, a certain method to find their solution.
2. **Ill-defined problems.** These are often called combinatorially explosive—more on what this means specifically later—"insight" problems that require a novel approach. These problems

are complex, and there is a way to find a solution, but the method has yet to be delineated.
3. **Undefinable problems.** This category is a special set. They cannot be "solved" because they comprise universal existential constraints for life on Earth, the inherent chaos of never being absolutely sure about a future interaction.

There is a core triplet of undefinable, unframeable problems.

They are the three descriptions of our finitude, one for each of the landscapes or planes of perception. In other words:

1. All life dies (the unsolvable final truth for the on-the-surface survival landscape).
2. Each life has limited power to forestall death (the unsolvable final truth for the above-the-surface thriving landscape).
3. No single life can command and control the universe (the unsolvable final truth for the beyond-the-surface deriving landscape).

So, from the Latin definition, we can conceptualize "intelligence" as a three-step problem-solving process:

- Step One: Read
- Step Two: Gather
- Steph Three: Choose

that manages reading, choosing, and gathering on three landscapes simultaneously:

- Landscape One: On-the-surface Survive
- Landscape Two: Above-the-surface Thrive
- Landscape Three: Beyond-the-surface Derive

Each of these landscapes have three kinds of problems within, they are:

- Problem Set One: Well-defined problems
- Problem Set Two: Ill-defined problems
- Problem Set Three: Undefinable problems

And we know the big kahuna problem is from problem set three for each of the landscapes, respectively.

- Landscape One's Undefinable Problem: No one will ever be absolutely alive or absolutely dead.
- Landscape Two's Undefinable Problem: No one will ever be absolutely powerful or absolutely impotent.
- Landscape Three's Undefinable Problem: No one will ever be in absolute command or be absolutely controlled.

Can you see why it's important to carefully consider how wonderfully complicated the word "intelligence" is?

REFERENCES

B-52's, vocalists. 1980. "Private Idaho." By Fred Schneider, Keith Strickland, Ricky Wilson, Cindy Wilson, and Kate Pierson. Track 5 on *Wild Planet*. Warner Bros.

Baum, L. Frank. [1900] 2021. *The Wonderful Wizard of Oz*. New York: Harper Design.

Bhattacharya, Ananyo. 2022. *The Man from the Future: The Visionary Life of John Von Neumann*. New York: W.W. Norton & Company.

Boole, George. 1958. *An Investigation of the Laws of Thought: On Which Are Founded the Mathematical Theories of Logic and Probabilities*. New York: Dover Publications.

Cohen, Leonard, vocalist. 1984. "Hallelujah." By Leonard Cohen. Side 2 track 1 on *Various Positions*. Columbia.

Copernicus Nicolaus. [1543] 1995. *On the Revolutions of Heavenly Spheres*. Amherst N.Y: Prometheus Books.

Dam, H. J. W. 1896. "The New Marvel in Photography" *McClure's Magazine*. VI, no. 5 (April 1896). https://www.radhis.nl/uploads/5/5/4/1/55414293/dam_1896.pdf.

Descartes, René. 1641. *Meditations on First Philosophy*. https://www.gutenberg.org/ebooks/23306.

Einstein, Albert. 1920. *Relativity: The Special and General Theory*. 1st ed. London: Methuen.

Feynman, Richard P., Ralph Leighton, and Edward Hutchings. 1997. *Surely You're Joking Mr. Feynman!: (Adventures of a Curious Character)*. New York: W.W. Norton & Company.

Galilei, Galileo, Stillman Drake, Albert Einstein, and J. L. Heilbron. [1632] 2001. *Dialogue Concerning the Two Chief World Systems Ptolemaic and Copernican*. New York: Modern Library.

Geisel, Theodor [Dr. Seuss, pseud.]. 1971. *The Lorax*. New York: Random House.

Gödel, Kurt, and Bernard Meltzer, translator. [1931] 1992. *On Formally Undecidable Propositions of Principia Mathematica and Related Systems*. New York: Dover Publications.

Hayes, Patrick J., and Leora Morgenstern. 2007. *AI Magazine* 28 no. 4 (Winter): 93–102.

Heaven, Will Douglas. 2023. "Geoffrey Hinton Tells Us Why He's Now

Scared of the Tech He Helped Build." *MIT Technology Review*. May 2, 2023. https://www.technologyreview.com/2023/05/02/1072528/geoffrey-hinton-google-why-scared-ai/.

Heidegger, Martin, John Macquarrie, and Edward S Robinson. (1927) 2008. *Being and Time*. New York: Harper Perennial/Modern Thought.

Hinton, Geoffrey. 2023. "Geoffrey Hinton Talks about the 'Existential Threat' of AI." interviewed by Will Douglas Heaven at EmTech Digital. May 3, 2023. https://www.technologyreview.com/2023/05/03/1072589/video-geoffrey-hinton-google-ai-risk-ethics/.

Hinton, Geoffrey, Peter Dayan, Brendan J. Frey, Radford M. Neal. 1995. "The Wake-Sleep Algorithm for Unsupervised Neural Networks." Department of Computer Science, University of Toronto. April 3,1995. https://www.cs.toronto.edu/~hinton/absps/ws.pdf.

Hodges, Andrew. [1983] 2014. *Alan Turing: The Enigma: The Book That Inspired the Film The Imitation Game*. Princeton, New Jersey: Princeton University Press.

Hürter, Tobias, and David Shaw, translator. 2022. *Too Big for a Single Mind: How the Greatest Generation of Physicists Uncovered the Quantum World*. New York: The Experiment.

Jackson, Peter. 2011. *The Lord of the Rings the Fellowship of the Ring*. New Line Home Entertainment. My Book.

Jeffress, Lloyd Alexander, editor. 1951. *Cerebral Mechanisms in Behavior: The Hixon Symposium*. New York: Wiley & Sons.

Jungk, Robert and James Cleugh. 1958. *Brighter Than a Thousand Suns: A Personal History of the Atomic Scientists*. New York: Harcourt Brace.

Kant, Immanuel and Mary J Gregor, translator. [1785] 1998. *Groundwork of the Metaphysics of Morals*. Cambridge U.K: Cambridge University Press.

Kline, Ronald R. 2011. "Cybernetics, Automata Studies, and the Dartmouth Conference on Artificial Intelligence." *IEEE Annals of the History of Computing*. 33, no. 4: 5–16. April 2011. doi: 10.1109/MAHC.2010.44.

Köhler, Wolfgang. [1970] 1992. *Gestalt Psychology: The Definitive Statement of the Gestalt Theory*. New York: Liveright.

Kondepudi, Dilip, and Ilya Prigogine. 1998. *Modern Thermodynamics: From Heat Engines to Dissipative Structures*. New York: John Wiley.

Laplace, Pierre-Simon, Frederick Wilson Truscott, and Frederick Lincoln Emory translators. 1814. "A Philosophical Essay on Probabilities." https://bayes.wustl.edu/Manual/laplace_A_philosophical_essay_on_probabilities.pdf

Lee, Peggy, vocalist. 1969. "Is That All There Is?" By Jerry Leiber and Mike Stoller. Track 1 on *Is That All There Is?* Capitol.

Levinson, Mark. 2018. *The Bit Player*. Institute of Electrical and Electronics Engineers (IEEE). My Book.

McCarthy, J., M. L. Minsky, N. Rochester, C.E. Shannon. 1955. "A Proposal for the Dartmouth Summer Research Project on Artificial Intelligence." August 31, 1955. reprinted in *AI Magazine*. 27: 12–14 (Winter 2006). http://jmc.stanford.edu/articles/dartmouth/dartmouth.pdf.

Melville, Herman. [1853] 2021. *Bartleby the Scrivener: A Story of Wall Street*. Mint Editions.

Nagel, Thomas. 1974. "What Is It Like to Be a Bat?" *The Philosophical Review*. 83, no. 4 (October 1974): 435–450.

Nilsson, Nils J. 2012. "John McCarthy 1927–2011" in *Biographical Memoirs National Academy of Sciences*. http://www.nasonline.org/publications/biographical-memoirs/memoir-pdfs/mccarthy-john.pdf.

O'Neill, Eugene. 1956. *Long Day's Journey into Night*. First ed. New Haven: Yale University Press.

Penn, Jonathan. 2021. "Inventing Intelligence: On the History of Complex Information Processing and Artificial Intelligence in the United States in the Mid-Twentieth Century." Apollo—University of Cambridge Repository. doi:10.17863/CAM.63087.

Portis Charles. [1968] 2004. *True Grit: A Novel*. New York: Overlook Press.

Prigogine, Ilya and Isabelle Stengers. [1978] 2018. *Order Out of Chaos: Man's New Dialogue with Nature*. New York: Verso.

R.E.M., vocalists. 1987. "It's the End of the World as We Know It (And I Feel Fine)." By Bill Berry, Peter Buck, Mike Mills, and Michael Stipe. Track 6 on *Document*. Sound Emporium.

Ramis, Harold, director. 1993. *Groundhog Day*. Columbia Pictures. My Book.

Rhodes, Richard. 1995. *Dark Sun: The Making of the Hydrogen Bomb*. New York: Simon & Schuster.

Sartre, Jean-Paul, and Sarah Richmond, translator. [1943] 2021. *Being and Nothingness: An Essay on Phenomenological Ontology*. New York: Atria.

Scott-Heron, Gil. 1971. "The Revolution Will Not Be Televised." By Scott-Heron. Track 1 on *Pieces of a Man*. RCA Studios.

Shannon, Claude E. 1948. "A Mathematical Theory of Communication," *The Bell System Technical Journal*, 27: 379–423, 623–656, July, October 1948. https://people.math.harvard.edu/~ctm/home/text/others/shannon/entropy/entropy.pdf.

Shannon, Claude E., and Warren Weaver. [1949] 1971. *The Mathematical Theory of Communication*. Urbana, Illinois: Univ. of Illinois Press.

Shannon, Claude Elwood, and John McCarthy. [1956] 1965. *Automata Studies*. Princeton, New Jersey: Princeton University Press.

Shannon, Claude. 1937. "A Symbolic Analysis of Relay and Switching Circuits," published in *Transactions of the American Institute of Electrical Engineers*. 57 no. 12: 713–723 (December 1938). https://www.cs.virginia.edu/~evans/greatworks/shannon38.pdf.

Skinner, B. F. 1938. *The Behavior of Organisms: An Experimental Analysis*. New York: D. Appleton-Century Company Incorporated.

Solomonoff, Grace. 2023. "The Meeting of the Minds That Launched AI." *IEEE Spectrum*. May 6, 2023. https://spectrum.ieee.org/dartmouth-ai-workshop.

Soni, Jimmy, and Rob Goodman. 2017. *A Mind at Play: How Claude Shannon Invented the Information Age*. New York: Simon & Schuster.

Von Neumann, John. 1993. "First draft of a Report on the EDVAC." *IEEE Annals of the History of Computing*. 15, no. 4: 27–75. doi: 10.1109/85.238389.

Watson, J., and F. Crick. 1953. "Molecular Structure of Nucleic Acids: A Structure for Deoxyribose Nucleic Acid." *Nature* 171: 737–38. https://doi.org/10.1038/171737a0.

Watson, James D. [1968] 2010. *The Double Helix: The Discovery of the Structure of DNA*. London: Phoenix.

Wiener Norbert. [1948] 2011. *Cybernetics: Or Control and Communication in the Animal and the Machine*. Cambridge Massachusetts: MIT

Zachary, G. Pascal. 1997. *Endless Frontier: Vannevar Bush, Engineer of the American Century*. New York: Free Press.

ABOUT JOHN VERVAEKE, PH.D.

John is an award-winning professor at the University of Toronto in the departments of psychology, cognitive science, and Buddhist psychology.

He currently teaches courses in the psychology department on thinking and reasoning with an emphasis on insight problem-solving, cognitive development with a focus on the dynamical nature of development, and higher cognitive processes with an emphasis on intelligence, rationality, mindfulness, and the psychology of wisdom.

He is the director of the cognitive science program where he also teaches courses on the introduction to cognitive science and the cognitive science of consciousness wherein he emphasizes 4E (embodied, embedded, enacted, and extended) models of cognition and consciousness.

In addition, he taught a course in the Buddhism, psychology, and mental health program on Buddhism and cognitive science for fifteen years. He is the director of the Consciousness and the Wisdom Studies Laboratory. He has won and been nominated for several teaching awards including the 2001 Students' Administrative Council and Association of Part-Time Undergraduate Students Teaching Award for the Humanities, and the 2012 Ranjini Ghosh Excellence in Teaching Award.

He has published articles on relevance realization, general intelligence, mindfulness, flow, metaphor, and wisdom. He is the first author of the book *Zombies in*

Western Culture: A Twenty-First Century Crisis, which integrates psychology and cognitive science to address the meaning crisis in Western society. He is the author and presenter of the YouTube series, "Awakening from the Meaning Crisis," "After Socrates" and the host of "Voices with Vervaeke."

ABOUT SHAWN COYNE

Shawn Coyne is a writer, editor, and publishing professional with over 30 years of experience. He has analyzed, acquired, edited, written, marketed, represented, or published 374 books with many dozens of bestsellers across all genres, and generated over $150,000,000 of revenue.

He graduated in 1986 with a degree in Biology from Harvard College, with a distinction of Magna Cum Laude for his thesis laboratory research work at the Charles A. Dana Laboratory of Toxicology at the Harvard School of Public Health. After Coyne left the laboratory, his findings were acknowledged and served as the inspiration for Mandana Sassanfar and Leona Samson's *Identification and Preliminary Characterization of an 06-Methylguanine DNA Repair Methyltransferase in the Yeast Saccharomyces cerevisiae* publication in the venerable *The Journal of Biological Chemistry* (Vol. 265, No. 1, Issue of January 5, pp. 20-25, 1990).

In 1991, early in his publishing career, Coyne began an independent investigation into the structure, function and organization of narrative, which he has since coined Simulation Synthesis Theory. His synoptic integration of Aristotle's *Poetics*, Freytag's *The Technique of the Drama*, Campbell's *Hero with a Thousand Faces*, McKee's *Story*, among many other story structure investigations with contemporary cognitive science, quantum information theory, cybernetics, evolutionary theory, behavioral psychology, Peircean and Jamesian pragmatism, Jungian depth psychology, Theologian and Philosopher Paul Tillich's conception of "ultimate concern," and fighter pilot John Boyd's OODA loop serves as philosophical, scientific and spiritual foundations for his teaching.

In 2015, he created *Story Grid Methodology* to begin teaching and further developing Simulation Synthesis Theory. Since then he has given lectures on the origin of story, the integration of storytelling and science, and the necessity of telling complex stories to thousands of students all over the world.

In addition to *The Story Grid* and *Mentoring the Machines,* he's authored, coauthored or ghost-written numerous bestselling nonfiction and fiction titles. His most recent lecture series, "Genre Blueprint" applies his Simulation Synthesis Theory to popular works such as *The Hobbit* by J.R.R. Tolkien and *The Matrix* by Lara and Lana Wachowski.